Revise IGCSE

Complete Study
&
Revision Guide

Bob McDuell

Chemistry

FOUNDATION®
BOOKS

Contents

8 Chemical reactions

9 Sulphur

10 Chemical changes

11 Acids, bases and salts

12 Metals

13 Electrochemistry and electrolysis

14 Experimental techniques

Preparing for the examination

Planning your study

The final three months before taking your IGCSE examination are very important in achieving your best grade. However, the success can be assisted by an organised approach throughout the course.

- After completing a topic in school or college, go through the topic again in Letts IGCSE Chemistry Guide. Copy the main points again on a sheet of paper or use a highlighter pen to emphasise them.
- A couple of days later try to write out these key points from memory. Check differences between what you wrote originally and what you wrote later.
- If you have written your notes on a piece of paper, keep this for revision later.
- Try some questions in the book and check your answers.
- Decide whether you have fully mastered the topic and write down any weaknesses you think you have.

Preparing a revision programme

Atleast three months before the final examination, go through the list of topics in your Examination Board's specification. Identify the topics which you feel you need to concentrate on. It is a temptation at this time to spend valuable revision time on the things you already know and can do. It makes you feel good but does not move you forward.

When you feel you have mastered all the topics, spend time trying past questions. Each time check your answers with the answers given. In the final couple of weeks, go back to your summary sheets (or highlighting in the book).

How this book will help you

Letts IGCSE Chemistry Guide will help you because:

- it contains the **essential content** for your IGCSE course without the extra material that will not be examined
- it contains **Progress checks** and **IGCSE questions** to help you to enhance your understanding
- it gives **sample IGCSE questions** with answers and advice from an examiner on how to improve your answers
- **marginal comments** and **highlighted key points** to draw your attention to important things you might otherwise miss.

Five ways to improve your grade

1. Read the question carefully

Many students fail to answer the actual question set. Perhaps they misread the question or answer a similar question they have seen before. Read the question once right through and then again more slowly. Some students underline or highlight key words in the question, as they read it through. Questions at IGCSE contain a lot of information. You should be concerned if you are not using that information in your answer.

2. Give enough detail

If a part of a question is worth three marks, you should make atleast three separate points. Be careful that you do not reiterate the same point thrice. Approximately 25% of the marks on your final examination papers are awarded for questions requiring longer answers.

3. Quality of Written Communication (QWC)

From 2003 onwards some marks on IGCSE papers are given for the quality of your written communication. This includes correct sentence structures, correct sequencing of events and use of scientific words.
Read your answer slowly before moving on to the next part.

4. Correct use of scientific language

There is important scientific vocabulary which should be used. Try to use the correct scientific terms in your answers and spell them correctly. The way scientific language is used is often a difference between successful and unsuccessful students. As you revise, make a list of scientific terms you meet and check that you understand the meanings of these words.

5. Show your working

All Chemistry papers include calculations. You should always show your complete working. Then, if you make an arithmetical mistake, you may still receive marks for correct procedure. Check that your answer is given to the correct number of significant figures and includes the correct unit.

Chapter 1

The particulate nature of matter

The following topics are covered in this section:

- ***The states of matter***
- ***Kinetic particle theory***
- ***Diffusion***

LEARNING SUMMARY

After studying this section you should be able to:

- ***describe the states of matter***
- ***explain kinetic particle theory***
- ***explain the behaviour of different states of matter in terms of the kinetic theory***
- ***describe and explain diffusion.***

Matter can be defined as a physical substance having mass and occupying space. All the physical things in the universe are composed of matter.

The states of matter

Some rare substances e.g. liquid crystals, used in flat panel displays, do not fall in either of the three states

Matter exists mostly in three states viz. solid, liquid and gas.

Solid state is one in which, under constant pressure and temperature, matter has a definite volume and shape.

Liquid state is one in which, under constant pressure and temperature, matter has a definite volume but not a definite shape. The shape varies with the shape of the container in which it is kept.

Gaseous state is one in which, even under constant pressure and temperature, matter has neither a fixed volume nor a fixed shape. It not only takes the shape of the container in which it is placed but it also spreads evenly to each and every corner of the container.

Kinetic particle theory

Kinetic theory of matter states that all types of matter are composed of tiny particles which are in constant motion, the physical properties of matter are dependent on the movement of these constituent particles.

The main points of kinetic particle theory are listed below:

- all matter is made up of very small particles, invisible to the naked eye, which are in continuous motion. Different substances are made up of different types of particles (e.g. atoms, molecules or ions), which have different sizes.
- The motion of these constituent particles is dependent on temperature and pressure – the more the temperature,the faster is the movement of these particles. Similarly, the more the pressure on them, the less is the movement in them.
- The motion of these particles is also dependent on their own weight – the lighter particles move at higher speeds than the heavier ones, at a given temperature.

Different states of matter in terms of kinetic particle theory

The kinetic theory can be used as a scientific model to explain the different states of matter and the interrelationship between them.

According to this theory, the behaviour of solids,liquids and gases can be explained as follows:

- In **solids**, the constituent particles are held close together by interparticle attractive forces, the particles have little freedom of movement and they can only vibrate about a fixed position.
- In **liquids**, the constituent particles are still close together but the forces of attraction are weaker and the particles have more freedom of movement. The particles have more energy and they move around in a random manner, often colliding with one another.
- In **gases**, the forces of attraction between the constituent particles is so weak and their individual energy is so high that forces of attraction become ineffective and the individual particles are free to move around, restricted only by the walls of the container they are kept in.The particles move randomly at very high speeds, colliding with each other and with the walls of the container in the process.

The behaviour of different states of matter in terms of the kinetic theory is summarised in **Table 1.1**.

Table 1.1 Different states of matter

State	Shape	Density	Interparticle forces	Particle movement	Particle seperation	Illustration
SOLID	*Fixed-has a definite shape*	*Fixed and quiet high*	*Forces of attraction greater than average energy of particles*	*Vibrate around fixed point*	*Particles held close together*	00000 00000 00000

(continued...)

Table 1.1 Different states of matter *(continued...)*

State	Shape	Density	Interparticle forces	Particle Movement	Particle Seperation	Illustration
LIQUID	*Variable- fills the container to a given level*	*Fixed and quiet high*	*Forces of attraction comparable to average energy of particles*	*Continuous random motion*	*Close but slightly seperated*	0 0 0 0 0 0
GAS	*Variable- totally fills container*	*Variable and quite low*	*Forces of attraction much less than average energy of particles*	*Continuous random motion– seperated collisions*	*Widely seperated*	0 0 0

Interconversion of different states of matter in terms of kinetic particle theory

The kinetic theory can be used to explain how a substance undergoes change from one state to another.

If energy of particles of a solid is increased, say by applying heat, the attractive forces binding the particles together are weakened and the particles are able to move around more freely, resulting in a flexible shape and the solid state gradually changes to a **liquid** one.The temperature at which this transition from solid to liquid starts is called the **melting point** of the solid.

Now, if this **liquid** is further heated, the enegy of the particles is increased even further, resulting in faster movement and subsequent thinning of the liquid. Finally, a stage is reached when the particles at the surface acquire enough energy to overcome the forces of attraction holding them together and they escape to form a **gas**.The temperature at which this transition from liquid to gas starts is called the **boiling point** of the liquid and the process of conversion of liquid to gas is called **evaporation**.

This process of change of solid state to liquid and then to gas can be totally reversed by decreasing the temperature of the gas, which results in decrease in average energy of the gas particles, making the forces of attraction stronger. As a result, the gas starts to **condense** into a liquid. If this liquid is further cooled, the energy of particles is decreased to such an extent that it is fully overcome by the attractive forces and the liquid **freezes** to become a solid.

Thus, it is amply clear that matter can change from one physical state to another and this process can also be reversed i.e. matter can be **interconverted** into different states by changing the external environment.

Diffusion–evidence for moving particles

Diffusion is the process of spreading out of particles when they move from a region of higher concentration to one of lower concentration. It is observed in liquids and gases but not in solids.

- Diffusion provides concrete evidence of particle movement since without particle movement, diffusion would not be possible.
- Diffusion is not observed in solids because individual movement of particles is severely restricted in the case of solids.
- The process of diffusion is slower in liquids as compared to gases because (i) the energy of constituent particles is less in liquid state and (ii) due to close packing of particles in liquids, free movement of diffusing particles is hindered.
- The **rate of diffusion** is dependent on molecular weight of diffusing matter – the lesser the weight of constituent particles, the higher is their speed of movement and hence higher is the rate of diffusion.
- **Evidence for moving particles in gases** can be observed by spraying perfume in one corner of a room, after some time we can smell the same scent in other corners of the room too, where the perfume vapours have moved (**diffused**) to.
- **Evidence for moving particles in liquids** can be observed by pouring a few drops of ink in a glass jar, filled with water. On observing the jar after a few hours, one can see that the ink particles have spread throughout and all the water in the container has become coloured.

Sample IGCSE questions

1. Everyday experiences can often be explained using scientific ideas about the way particles are arranged in solids, liquids and gases.

(a) Show the arrangement of particles in a typical solid, liquid and gas (in each case, use the symbol X to represent each particle). **[3]**

Refer to pages 9-10, Table 1.1.

(b) When a jar of coffee is opened, people in all parts of the room soon notice the smell. Use ideas about particles to explain this phenomena. **[2]**

Answer should include something about diffusion of particles.

2. Why do solids get converted to liquids when heated? **[2]**

Soilds get converted to liquids as the particles gain energy due to heating. The attractive forces binding the particles together are weakened and the particles are able to move around more freely, resulting in a flexible shape and the solid state gradually changes to liquid.

Exam practice questions

1. Salt soon disappears when stirred in water. Explain then why all parts of the water taste salty. **[2]**

2. When we use perfume, why can it be smelled from a distance? **[2]**

3. Why do liquids turn to solids on cooling? **[2]**

Chapter 2

Atoms, elements and compounds

The following topics are covered in this section:

- ***Atomic structure***
- ***Bonding***

2.1 Atomic structure and periodic table

LEARNING SUMMARY

After studying this section you should be able to:

- ***recall the particles that make up all atoms and the properties of these particles***
- ***work out the numbers of protons, neutrons and electrons using mass number and atomic number***
- ***understand the relationship between the position of an element in the periodic table and the properties of the element***
- ***explain the relationship between the position of an element in the periodic table and the arrangement of electrons in the atoms***
- ***explain why some elements contain different isotopes.***

Particles in an atom

All **elements** are made up from **atoms**.

KEY POINT

An atom is the smallest part of an element that can exist.

About 2000 years ago, Democritus, a Greek philosopher, claimed that all substances were made up of atoms. He had no evidence for this so it was dismissed. Until the early 19th century, scientists believed that atoms were indivisible – like snooker balls. In 1803, Dalton revived the idea of matter being made up of atoms.

It has been found that the atoms of all elements are made up from three basic particles and that the atoms of different elements contain different numbers of these three particles. These particles are:

Particle	**Mass**	**Charge**
Proton p	*1 u (u is atomic mass unit)*	*+1*
Electron e	*Negligible*	*–1*
Neutron n	*1 u*	*Neutral*

KEY POINT

Because an atom has no overall charge, the number of protons in any atom is equal to the number of electrons.

In the atom the protons and neutrons are tightly packed together in the **nucleus**. The nucleus is **positively charged**. The electrons move around the nucleus in **energy levels** or **shells**.

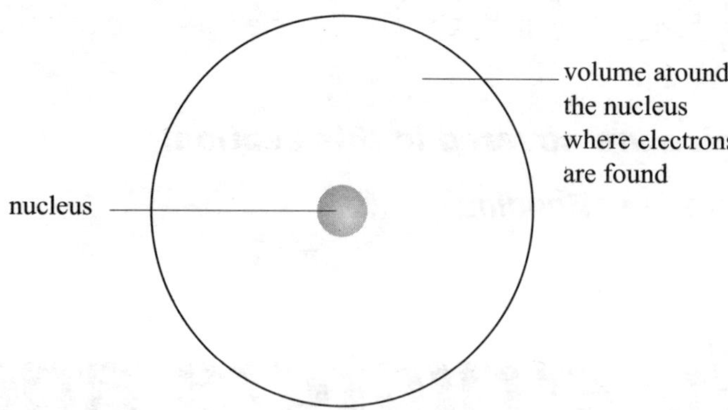

Fig 2.1 shows a simple representation of an atom.

Most of an atom is empty space. If a football stadium represented an atom, the nucleus would be the size of the centre spot.

Atomic number and mass number

There are two 'vital statistics' for any atom.

- **Atomic number**

The atomic number is the number of **protons** in an atom.

Candidates often get atomic number and mass number confused

- **Mass number**

The mass number is the total number of **protons and neutrons** in an atom.

Atomic number is sometimes called proton number.

We can use these numbers for any atom to work out the number of protons, neutrons and electrons.

Protons and neutrons are also called nucleons, as they reside in nucleus of the atom.

Mass number is also called **nucleon number**.

E.g. The mass number of carbon-12 is 12, and the atomic number is 6.

Therefore a carbon-12 atom contains 6 protons (i.e. atomic number = 6), 6 electrons and 6 neutrons. This is sometimes written as:

All atoms of the same element have the same atomic number and contain the same number of protons and electrons.

$$^{12}_{6}C$$

(the atomic number is written under the mass number).

For an atom of sodium-23:

Sodium -23 can be written: $^{23}_{11}Na$

mass number = 23; atomic number = 11

number of protons = 11

number of electrons = 11

number of neutrons = 23 – 11= 12.

Structure of the periodic table

The periodic table is an arrangement of all of the hundred plus elements in order of increasing atomic number, with elements with similar properties in the same vertical column.

The periodic table is shown in its modern form in **Fig. 2.2**.

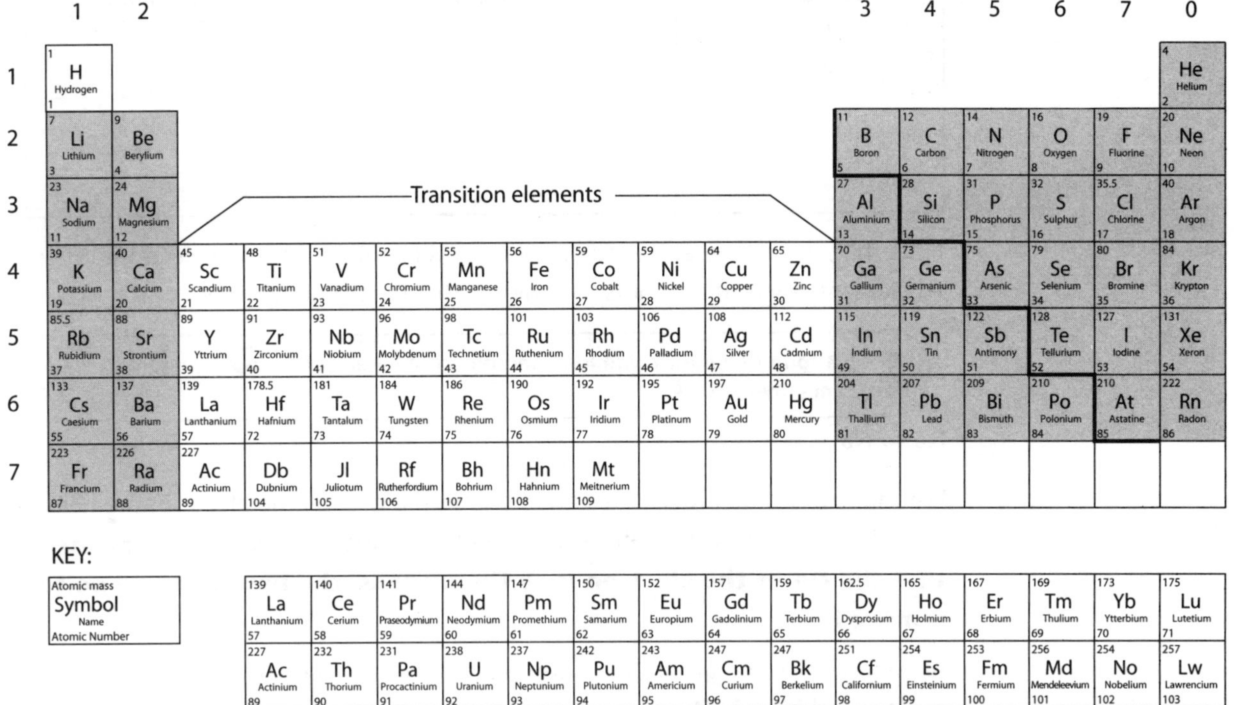

	1	2											3	4	5	6	7	0
1	1 H Hydrogen 1																	4 He Helium 2
2	7 Li Lithium 3	9 Be Berylium 4											11 B Boron 5	12 C Carbon 6	14 N Nitrogen 7	16 O Oxygen 8	19 F Fluorine 9	20 Ne Neon 10
3	23 Na Sodium 11	24 Mg Magnesium 12	Transition elements										27 Al Aluminium 13	28 Si Silicon 14	31 P Phosphorus 15	32 S Sulphur 16	35.5 Cl Chlorine 17	40 Ar Argon 18
4	39 K Potassium 19	40 Ca Calcium 20	45 Sc Scandium 21	48 Ti Titanium 22	51 V Vanadium 23	52 Cr Chromium 24	55 Mn Manganese 25	56 Fe Iron 26	59 Co Cobalt 27	59 Ni Nickel 28	64 Cu Copper 29	65 Zn Zinc 30	70 Ga Gallium 31	73 Ge Germanium 32	75 As Arsenic 33	79 Se Selenium 34	80 Br Bromine 35	84 Kr Krypton 36
5	85.5 Rb Rubidium 37	88 Sr Strontium 38	89 Y Yttrium 39	91 Zr Zirconium 40	93 Nb Niobium 41	96 Mo Molybdenum 42	98 Tc Technetium 43	101 Ru Ruthenium 44	103 Rh Rhodium 45	106 Pd Palladium 46	108 Ag Silver 47	112 Cd Cadmium 48	115 In Indium 49	119 Sn Tin 50	122 Sb Antimony 51	128 Te Tellurium 52	127 I Iodine 53	131 Xe Xeron 54
6	133 Cs Caesium 55	137 Ba Barium 56	139 La Lanthanium 57	178.5 Hf Hafnium 72	181 Ta Tantalum 73	184 W Tungsten 74	186 Re Rhenium 75	190 Os Osmium 76	192 Ir Iridium 77	195 Pt Platinum 78	197 Au Gold 79	210 Hg Mercury 80	204 Tl Thallium 81	207 Pb Lead 82	209 Bi Bismuth 83	210 Po Polonium 84	210 At Astatine 85	222 Rn Radon 86
7	223 Fr Francium 87	226 Ra Radium 88	227 Ac Actinium 89	Db Dubnium 104	Jl Juliotum 105	Rf Rutherfordium 106	Bh Bohrium 107	Hn Hahnium 108	Mt Meitnerium 109									

KEY:

Atomic mass Symbol Name Atomic Number

139 La Lanthanium 57	140 Ce Cerium 58	141 Pr Praseodymium 59	144 Nd Neodymium 60	147 Pm Promethium 61	150 Sm Samarium 62	152 Eu Europium 63	157 Gd Gadolinium 64	159 Tb Terbium 65	162.5 Dy Dysprosium 66	165 Ho Holmium 67	167 Er Erbium 68	169 Tm Thulium 69	173 Yb Ytterbium 70	175 Lu Lutetium 71
227 Ac Actinium 89	232 Th Thorium 90	231 Pa Procactinium 91	238 U Uranium 92	237 Np Neptunium 93	242 Pu Plutonium 94	243 Am Americium 95	247 Cm Curium 96	247 Bk Berkelium 97	251 Cf Californium 98	254 Es Einsteinium 99	253 Fm Fermium 100	256 Md Mendeleevium 101	254 No Nobelium 102	257 Lw Lawrencium 103

The vertical columns in the periodic table are called **groups**. **The elements in a group have similar properties**.

The horizontal rows of elements are called **periods**.

The **main block** of elements are shaded in **Fig. 2.2**. The elements between the two parts of the main block are the **transition metals**.

The bold stepped line on the table divide **metals** on the left hand side from **non-metals** on the right.

Relationship between electron arrangement and position in the periodic table

> **KEY POINT**
>
> **For any element in the main block of the periodic table, it is easy to work out the electron arrangement in the atoms.**
>
> - **The number of energy levels or shells is the same as the period in which the element is placed.**
> - **The number of electrons in the outer energy level is the same as the group number (except for elements in group 0 which have 8 electrons, apart from helium which has two electrons).**

Strontium is in period 5 and group 2. This means that five energy levels are used with two electrons in the outer energy level.

If you look up the electron arrangement of strontium, it is 2, 8, 18, 8, 2.

Table 2.1 shows the arrangement of electrons in atoms of alkali metals (group 1).

Element	Atomic number	Electron arrangement
Li	*3*	*2,1*
Na	*11*	*2,8,1*
K	*19*	*2,8,8,1*
Rb	*37*	*2,8,18,8,1*
Cs	*55*	*2,8,18,18,8,1*

Note that, in each case, the outer energy level contains just one electron. When an element reacts, it attempts to achieve a full outer energy level.

KEY POINT **Group I elements will lose one electron when they react and form a positive ion.**

$Na \rightarrow Na^{+} + e^{-}$

We can explain the order of reactivity within the group. The electrons are held in position by the electrostatic attraction of the positive nucleus. **This means that the closer the electron is to the nucleus, the harder it will be to remove it**.

As we go down the group, the outer electron gets further away from the nucleus and **hence, it becomes easier to take it way**. This means as we go down the group, the reactivity should increase.

Table 2.2 shows the arrangement of electrons in atoms of halogens (group 7).

Element	Atomic number	Electron arrangement
F	*9*	*2,7*
Cl	*17*	*2,8,7*
Br	*35*	*2,8,18,7*
I	*53*	*2,8,18,18,7*

Note that each member of the halogens (group 7) has seven electrons in the outer energy level. This is just one electron short of the full energy level. When halogens react, they gain an electron to complete that outer energy level. This will form a negative ion.

The fact that reactivity increases down group 1 but decreases down group 7 frequently leads to mistakes by students.

$Cl + e^{-} \rightarrow Cl^{-}$

As an electron is being gained in the reaction, the most reactive member of the family will be the one where the extra electron is closest to the nucleus,i.e. fluorine.

The reactivity decreases down the group.

The elements in a group tend to have similar physical and chemical properties because of their similar outer shell electron configuration.

Except for the noble gases, the more the number of electrons (>4) in the outer shell, the more non-metallic and the more reactive is the element. The most reactive non-metals only need to share/gain one or two electrons.

The most reactive metals have a low number of outer shell valence electrons (<= 3) which are easily lost by them, forming the metal ion in reaction. The most reactive metals only have 1 or 2 electrons in the outer shell.

Elements in the 'middle' of the periodic table, e.g. group 4 with 4 outer electrons, show non-metallic character. e.g. carbon.

So, in a period, we have metallic elements on the left and non-metallic elements on the right.

Isotopes

In cases of many elements, it is possible to have more than one type of atom.

For example, there are three types of oxygen atom:

oxygen-16 8p, 8e, 8n

oxygen-17 8p, 8e, 9n

oxygen-18 8p, 8e, l0n

These three different atoms contain 8 protons and 8 electrons. This determines that all atoms are oxygen atoms.

These different types of atom of the same element are called **isotopes.**

KEY POINT **Isotopes are atoms of the same element containing the same number of protons and electrons but different numbers of neutrons.**

Some elements, e.g. fluorine, have only one isotope, but others have different isotopes. Calcium, for example, contains six isotopes.

Isotopes of the same element have the **same chemical properties** but slightly **different physical properties**. There are two isotopes of chlorine — chlorine-35 and chlorine-37. An ordinary sample of chlorine contains approximately 75 per cent chlorine-35 and 25 per cent chlorine-37. This explains the fact that the relative atomic mass of chlorine is approximately 35.5. (The relative atomic mass of an element is the mass of an 'average atom' compared with the mass of a $^{12}_{6}C$ carbon atom.)

Most elements have both stable and radioactive isotopes. Radioactive isotopes of an element are commonly used as traces in medical, biological and industrial studies to gain information about physical and mechanical process. In geology and archaeology, radioactive isotopes are used to determine the age of a sample.

Radioactive isotopes

Radioactive isotopes are unstable because of the extra neutrons in their nuclei. They decay, emitting alpha, beta, and sometimes gamma rays. Such isotypes eventually reach stability in the form of non-radioactive isotopes of other chemical elements.

- Medical use: Cobalt-60 is used in radiotherapy treatment of cancer.
- Industrial use: Uranium-235 is used as a source of power in nuclear reactors.

Non-radioactive/stable isotopes

Stable isotopes are those isotopes that do not undergo radioactive decay. Thus, their nuclei are stable and their masses remain the same. However, they may themselves be the product of the decay of radioactive isotopes.

2.2 Bonding

LEARNING SUMMARY

After studying this section you should be able to:

- ***understand the difference between elements, compounds and mixtures***
- ***distinguish between metals and non-metals***
- ***recall that atoms are joined together by chemical bonds***
- ***understand that ionic bonding takes place when one or more electrons are completely transferred from a metal atom to a non-metal atom***
- ***understand that covalent bonding involves the sharing of pairs of electrons***
- ***describe the differences between giant and molecular structures***
- ***understand that some elements, e.g. carbon, can exist in different forms in the same state. These forms are called allotropes.***

The joining of atoms together is called bonding. An arrangement of particles bonded together is called a structure.

There are several types of bonding found in common chemicals.

Three methods of bonding atoms together are **ionic** bonding, **covalent** bonding and **metallic** bonding.

Difference between elements, compounds and mixtures

An element is a pure substance made up of only one type of atom.

The chemical bonding of two or more elements forms compounds. Compounds are usually very different from the individual elements from which they are formed. The elements in a compound can be separated by means of a chemical reaction only.

E.g. Chlorine is a green coloured gas and sodium is a soft, silvery metal. However when the two combine, they form sodium chloride (table salt) – a colourless crystalline material. Sodium cannot be regained back from salt by a physical

In a mixture, two or more substances (elements or compounds) are blended together. Unlike compounds, it involves no chemical change and each substance in a mixture retains its individual properties. The substances in a mixture can usually be separated quite easily by physical means, e.g. filtration, distillation etc. without a chemical reaction.

E.g. When sugar is dissolved in water, it forms a mixture. This mixture retains the properties of its constituents and tastes sweet. Sugar can easily be separated from this mixture by evaporating the water.

Metals and non-metals

KEY POINT

Elements are broadly classified into metals and non-metals on the basis of their physical and chemical characteristics.

- The metallic bonding present in metals is very strong due to which most metals, unlike most non-metals, have **high density and strength** with **high melting and boiling points**.
- Metals are **good conductors** of electricity and heat whereas most non-metals are poor conductors.
- Typical metals also have a silvery **metallic surface** on freshly fractured faces.
- Metals are **malleable** i.e. they (unlike most non metals) can be readily bent, pressed or hammered into desired shapes.

A few elements display both metallic and non-metallic characteristics and are called semi-metals or **metalloids**.

Properties of metals can be changed by mixing them physically with other metallic or non-metallic materials. Such a mixture is called an **alloy**. It is not a compound as no chemical reaction is involved in its manufacture.

Ionic (or electrovalent) bonding

KEY POINT **Ionic bonding involves a complete transfer of electrons from one atom to another.**

Two examples are given below:

- **Sodium chloride**

A sodium atom has an electron arrangement of 2,8,1 (i.e. one more electron than the stable electron arrangement of 2,8).

Sodium chloride is a compound formed from the reaction of a reactive metal (sodium) with a reactive non-metal (chlorine).

A chlorine atom has an electron arrangement of 2,8,7 (i.e. one electron less than the stable electron arrangement – 2,8,8).

Both the **ions** formed have **stable electron arrangements**. The ions are held together by **strong electrostatic forces**.

KEY POINT **Each sodium atom loses one electron to form a sodium ion Na^+. Each chlorine atom gains one electron and forms a chloride ion Cl^-.**

It is important to stress in your answers that there is a complete transfer of electrons in ionic bonding. Electrons go from the metal (sodium) to the non-metal (chlorine). A frequent mistake is to use terms such as atoms swapping electrons. This is wrong!

This process is summarised in **Fig. 2.3**.

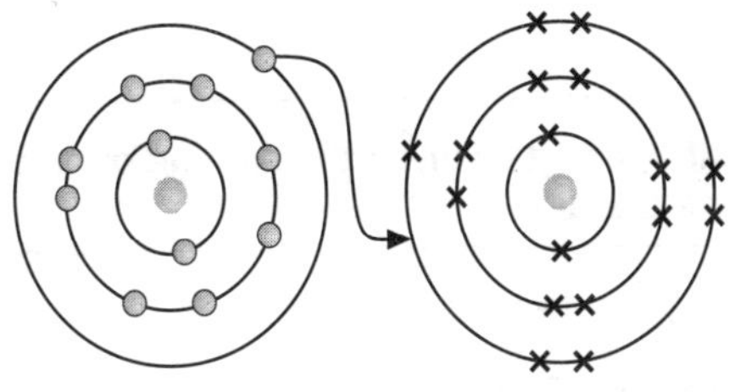

Fig. 2.3 Ionic bonding in sodium chloride

It is incorrect to speak of a 'sodium chloride **molecule**'. This would assume that one sodium ion joins with one chloride ion.

KEY POINT **A sodium chloride crystal consists of a regular arrangement of equal numbers of sodium and chloride ions. This is called a lattice.**

Fig 2.4 A sodium chloride lattice

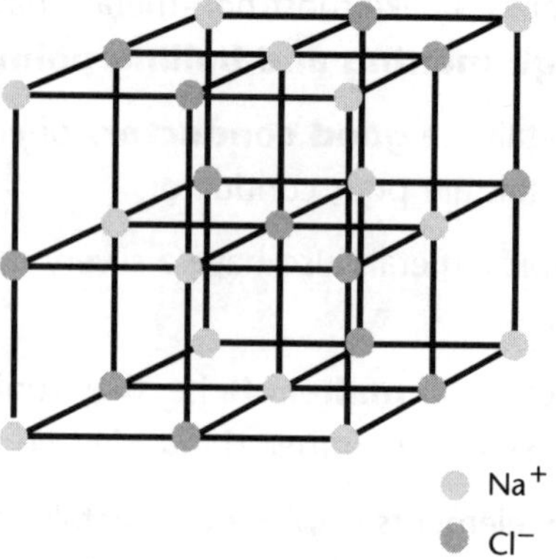

In ionic bonding, one element is a metal and one is a non-metal. Metal atoms lose electrons and non-metal atoms gain them.

- **Magnesium oxide**

Electron arrangement in magnesium atom 2,8,2.

Electron arrangement in oxygen atom 2,6.

KEY POINT Two electrons are lost by each magnesium atom to form Mg^{2+} ions. Two electrons are gained by each oxygen atom to form O^{2-} ions.

This is summarised in **Fig. 2.5.**

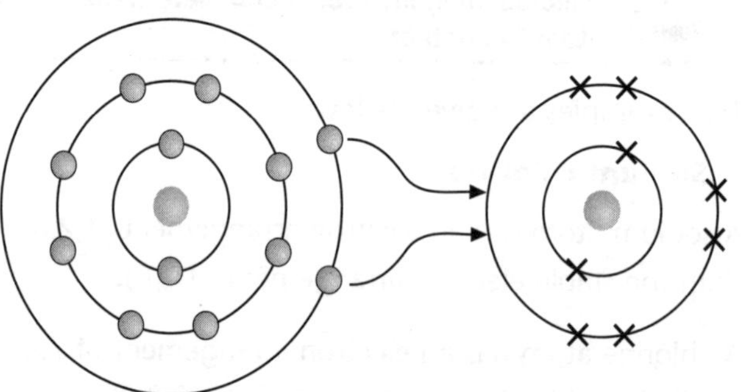

Fig. 2.5 Ionic bonding in magnesium oxide

2 electrons completely transferred

Loss of one or two electrons by a metal during ionic bonding is common, e.g. NaCl or MgO.

If three electrons are lost by a metal, the resulting compound shows some covalent character, e.g. $AlCl_3$.

Covalent bonding

KEY POINT Covalent bonding involves the sharing of electrons, rather than complete transfer.

Two examples are given below:

- **Chlorine molecule (Cl_2)**

A chlorine atom has an electron arrangement of 2,8,7. When two chlorine atoms bond together they form a chlorine **molecule**. If one electron was transferred from one chlorine atom to the other, only one atom could achieve a stable electron arrangement.

A similar covalent bond exists in a hydrogen molecule.

Instead, one electron from each atom is donated to form a **pair of electrons** which is shared between both atoms, holding them together. This is called a **single covalent** bond. **Fig. 2.6** shows a simple representation of a chlorine molecule using a dot and cross diagram.

This is often shown as Cl—Cl.

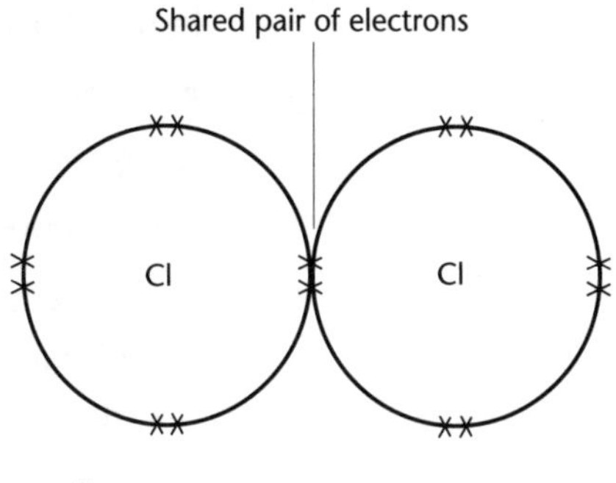

Fig. 2.6

- **Oxygen molecule (O_2)**

An oxygen atom has an electron arrangement of 2,6. In this case each oxygen atom donates two electrons and the **four electrons (two pairs)** are **shared** between both atoms. This is called a **double covalent bond**. **Fig. 2.6** shows a simplified representation of an oxygen molecule.

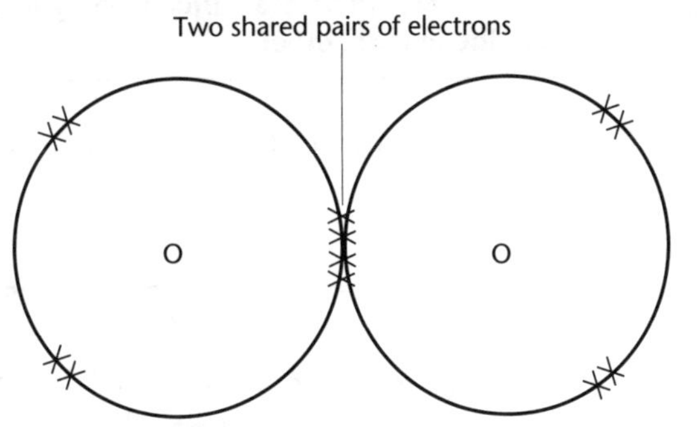

This is usually shown as O=O.

When drawing dot and cross diagrams, do not draw them too small. They can be difficult for the examiner to interpret. Remember both dots and crosses represent electrons.

Only electrons in the outer shell are drawn here. It makes the diagram simpler. Remember the inner shells are still there.

The figure below shows other examples of molecules containing covalent bonding.

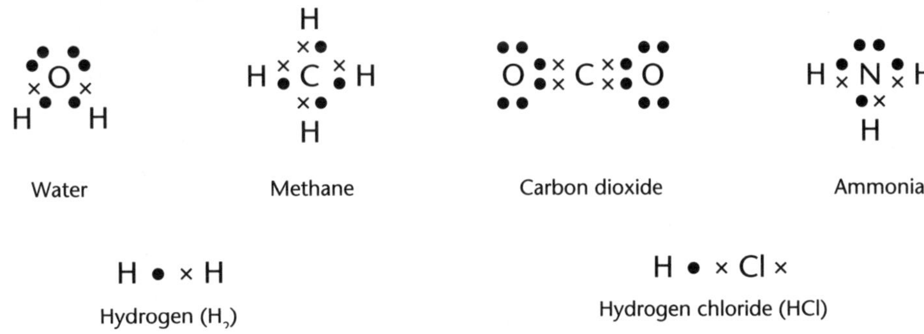

Fig. 2.7

Examples of some more complex covalent molecules are shown in **Fig. 2.8**.

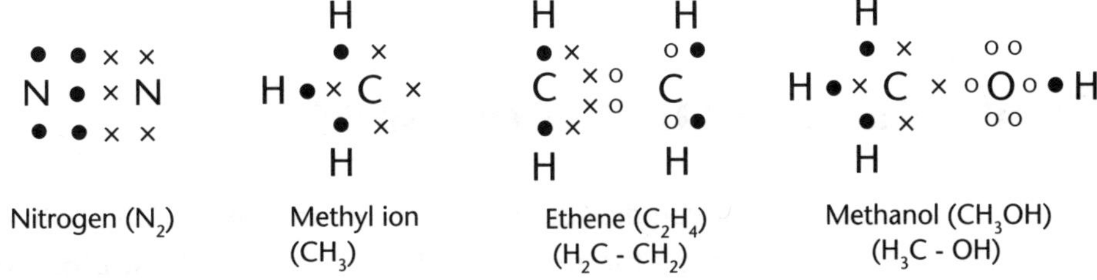

Fig. 2.8

In the above structures,

- nitrogen (N_2) has one triple bond.
- Methyl ion $-CH_3^+$ has 3 single bonds between C and H. It requires one more electron for C to complete 8 electrons.
- In ethene (C_2H_4), two atoms of carbon combine with four atoms of hydrogen. The structure has one carbon – carbon double bond and four carbon – hydrogen single bonds.
- Methanol (CH_3OH), has C from one $-CH_3$ group sharing its electron with O of –OH.

Metallic bonding

Metallic bonding is found only in metals.

A metal consists of a close-packed regular arrangement of positive ions, which are surrounded by a 'sea' of electrons that binds the ions together.

Fig. 2.9 shows the arrangement of ions in a single layer.

The sea of electrons can move throughout the structure. This explains the high electrical conductivity of solid metals. Metals are crystalline. This is due to the regular arrangement of particles in the structure.

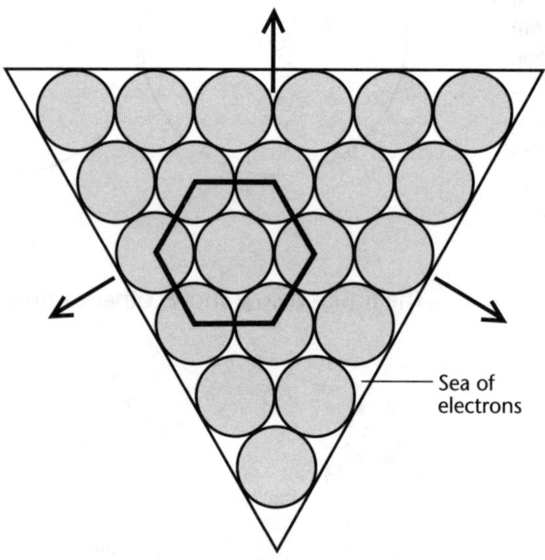

Fig. 2.9

There are two alternative ways of stacking these layers. The arrows in **Fig. 2.9** indicate that the layer shown continues in all directions. Around any one ion in a layer, there are six ions arranged in a hexagonal manner.

Metals are malleable, i.e., they can be readily hammered into any shape. The reason is that due to mobility of the electrons, the layers of atoms can slide over each other without fracturing the structure.

Giant and molecular structures

Silicon dioxide, SiO_2, and carbon dioxide, CO_2, both contain covalent bonding to join the atoms together. However, **silicon dioxide is a solid** and **carbon dioxide is a gas**.

In carbon dioxide, each carbon atom is joined with two oxygen atoms to form a **molecule**.

The molecules are not held together.

In silicon dioxide there is, in effect, one large molecule. Each silicon is bonded to four oxygen atoms and each oxygen is bonded to two silicon atoms. The resulting structure is called a **giant structure**.

In a molecular structure, there may be strong forces within each molecule but the forces between the molecules are very weak.

Fig. 2.10 shows simple representations of molecular and giant structures.

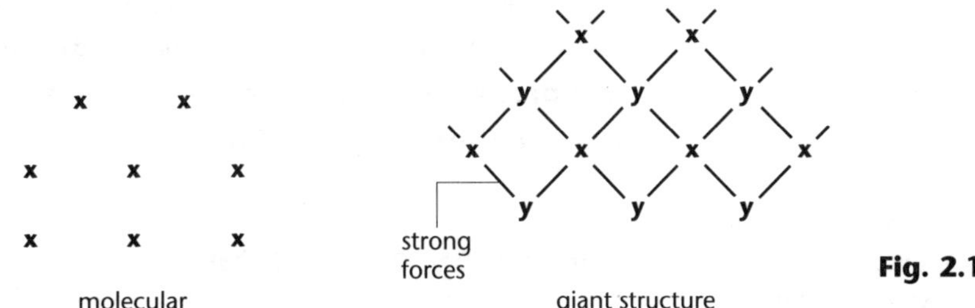

Fig. 2.10

Bonding, structure and properties

Table 2.3 below summarises how bonding and structure alter properties of substances.

Bonding	Structure	Properties
Ionic	**Giant structure,** *e.g. sodium chloride, magnesium oxide*	*High melting and boiling point, usually soluble in water but insoluble in organic solvents. Conduct electricity when molten or dissolved in water* **(electrolytes).**
Covalent	**Molecular,** *e.g. chlorine, iodine, methane*	*Usually gases or low boiling point liquids.Some (iodine and sulphur)are low melting point solids. Usually insoluble in water but soluble in organic solvents. Do not conduct electricity.*
	Macromolecules *(large molecules) e.g. poly(ethene), starch*	*Solids. Usually insoluble in water but more soluble in organic solvents. Do not conduct electricity.*
	Giant structure, *e.g. silicon dioxide*	*Solids. High melting points. Insoluble in water and organic solvents. Do not conduct electricity.*
Metallic	**Giant structure,** *e.g. copper*	*Solids. High density (ions closely packed). Good electrical conductors (free electrons).*

Solid sodium chloride does not conduct electricity. The ions are not free to move. In molten sodium chloride and sodium chloride solution, the ions are free to move and they conduct electricity.

Allotropy

Allotropy is the existence of two or more forms of an element in the same physical state.

Solid lead and molten lead are not allotropes because they are not in the same physical state – one is solid and the other is liquid.

These different forms are called **allotropes**. Allotropy is caused by the possibility of more than one arrangement of atoms. For example, carbon can exist in allotropic forms including **diamond** and **graphite**. Sulphur can exist in two allotropes — α-**sulphur** and β-**sulphur**.

Oxygen, O_2, and ozone, O_3, are two gaseous forms of oxygen. They are allotropes. You have probably heard of the ozone layer.

Allotropy of carbon

The two most commonly mentioned allotropes of carbon are diamond and graphite.

Another allotrope of carbon is **fullerene** which is a crystalline form of carbon made of clusters of carbon atoms.

1 Diamond

In the diamond structure, each carbon atom is strongly bonded (covalent bonding) to four other carbon atoms tetrahedrally

A large **giant structure** (three-dimensional) is built up. All bonds between carbon atoms are of the same length (0.154 nm). It is the strength and uniformity of the bonding which make diamond very hard, non-volatile and resistant to chemical attack. **Fig. 2.11** shows the arrangement of particles in diamond.

Other elements show allotropy including sulphur, phosphorus and tin.

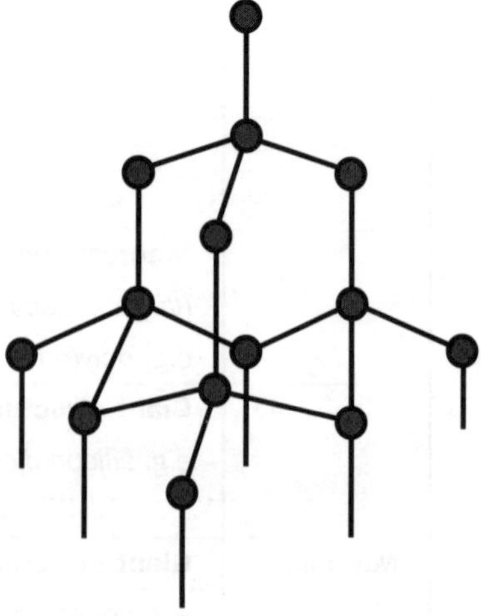

Fig. 2.11

Due to it's hardness, diamond is used as the 'leading edge' in cutting tools.

The structure of silicon dioxide is very similar to that of diamond, but it is nowhere near diamond, in terms of hardness, because the silicon-silicon bond is much weaker than the carbon-carbon bond present in diamond.

2 Graphite

Graphite has a **layered structure**. In each layer, the carbon atoms are bound covalently. The bonds **within the layers** are very **strong**.

> **KEY POINT** The bonds between the layers of graphite are very weak, which enables layers to slide over one another.

This makes the graphite soft and flaky.

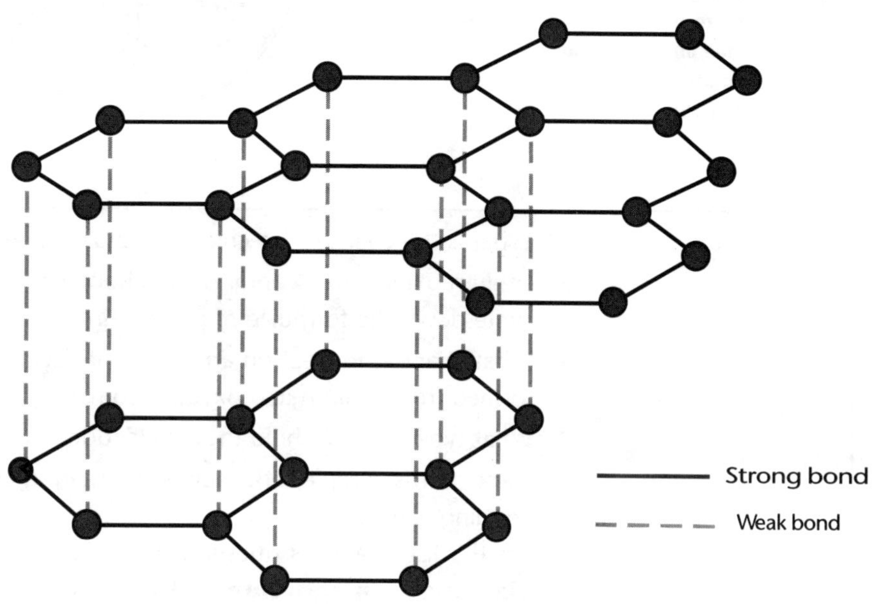

Fig. 2.12 Structure of graphite

> Due to its hardness, diamond is used as the 'leading edge in cutting tools.

It is the only non metal that is a good conductor of electricity, the reason being that electrons, from the 'shared bond', can move freely through each layer. Therefore, graphite is also used in electrical contacts, e.g. as electrodes in electrolysis.

Table 2.4 compares the properties of diamond and graphite.

Property	Diamond	Graphite
Appearance	*Transparent, colourless crystals*	*Black, opaque, shiny solid*
Density (g/cm³)	*3.5*	*2.2*
Hardness	*Very hard*	*Very soft*
Electrical conductivity	*Non-conductor*	*Good electrical conductor*

> Many other important discoveries have been made by accident, e.g. poly(ethene), xenon tetrafluoride.

A chance discovery in 1985 led to the identification of a new allotrope of carbon. In fact, a new family of closed carbon clusters has been identified and called **fullerenes**. Two fullerenes, C_{60} and C_{70}, can be prepared by electrically evaporating carbon electrodes in helium gas at low pressure. They dissolve in benzene to produce a red solution.

Fig. 2.13 shows a fullerene molecule.

Fullerenes are good lubricants as molecules can easily slide over each other.

The process for making fullerenes has to be carried out in atmosphere of helium. In air the carbon would burn.

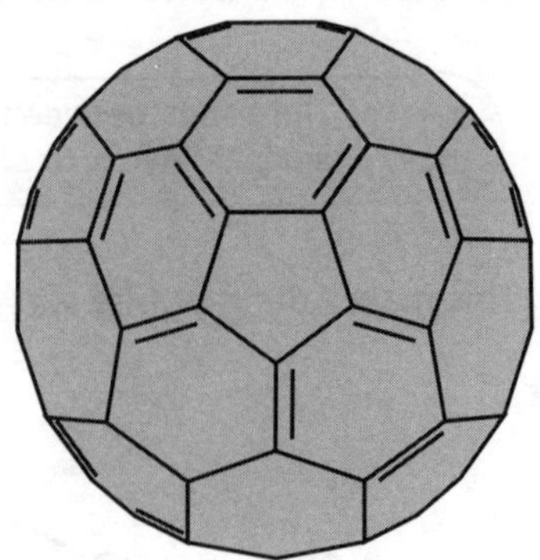

Fig. 2.13

PROGRESS CHECK

1. What is the charge on a chloride ion and how does this come about?
2. Lithium oxide, Li_2O, contains ionic bonding.
   Write down the formulae of the ions in lithium oxide.
3. What changes in electron arrangement occur when these ions are formed from lithium and oxygen atoms?
4. What type of forces hold these ions together in the solid?
5. Some atoms complete their shells by sharing electrons. What type of bonding is this?

Use this list to answer questions 6–9.

diamond **magnesium oxide** **methane** **silicon dioxide**

6. Which substance in the list is an example of an element with a giant structure of atoms?
7. Which substance in the list is an example of a giant structure of ions?
8. Which substance in the list is an example of a compound with a giant structure of atoms?
9. Which substance in the list is an example of a molecular structure?
10. Why are metals good conductors of electricity?

1. One negative charge – gains one electron; 2. Li^+ and O^{2-}; 3. Lithium atom loses one electron, oxygen atom gains two electrons; 4. Electrostatic; 5. Covalent; 6. Diamond; 7. Magnesium oxide; 8. Silicon dioxide; 9. Methane; 10. Free electrons move through the metal.

Sample IGCSE questions

1. Europium is an element discovered in 1901 by *E.A. Demarcay* in France.

It is a silvery-white metal.
It was given the symbol Eu.
Its atoms have an electron arrangement 2, 8, 18, 25, 8, 2.

(a) What is the atomic number of europium? **[1]**

63 ✓

Just count up the number of the electrons. This is the same as the number of protons and is equal to the atomic number.

(b) There are two isotopes of europium. They are europium-151 and europium-153. In a sample of europium there is 50% of each isotope.

(i) How many neutrons are there in each isotope? **[2]**

Europium–151 88 neutrons ✓
Europium–153 90 neutrons ✓

You can work out the number of neutrons by subtracting the atomic number from the mass number.

(ii) What is the relative atomic mass of europium? Explain your answer. **[2]**

152 ✓
Half way between 151 and 153 because the isotopes are present in equal amounts ✓.

(iii) How many particles are present in the nucleus of a europium-151 atom? **[1]**

151 — protons plus neutrons ✓

The particles in the nucleus are sometimes called nucleons.

2. The table shows the number of protons and electrons in sodium and fluorine atoms.

Atom	Number of protons	Number of electrons
Sodium	*11*	*11*
Fluorine	*9*	*9*

(a) Draw diagrams to show the arrangement of electrons in a sodium atom and in a fluorine atom. **[2]**

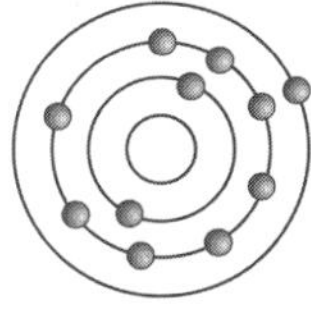

sodium ✓

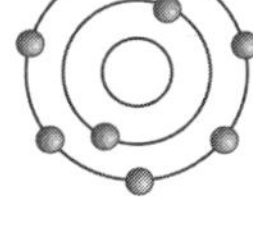

fluorine ✓

The periodic table could help you here.

Sample IGCSE questions

(b) **(i)** Draw a diagram to show outer electrons in a fluorine molecule, F_2. **[2]**

✓✓

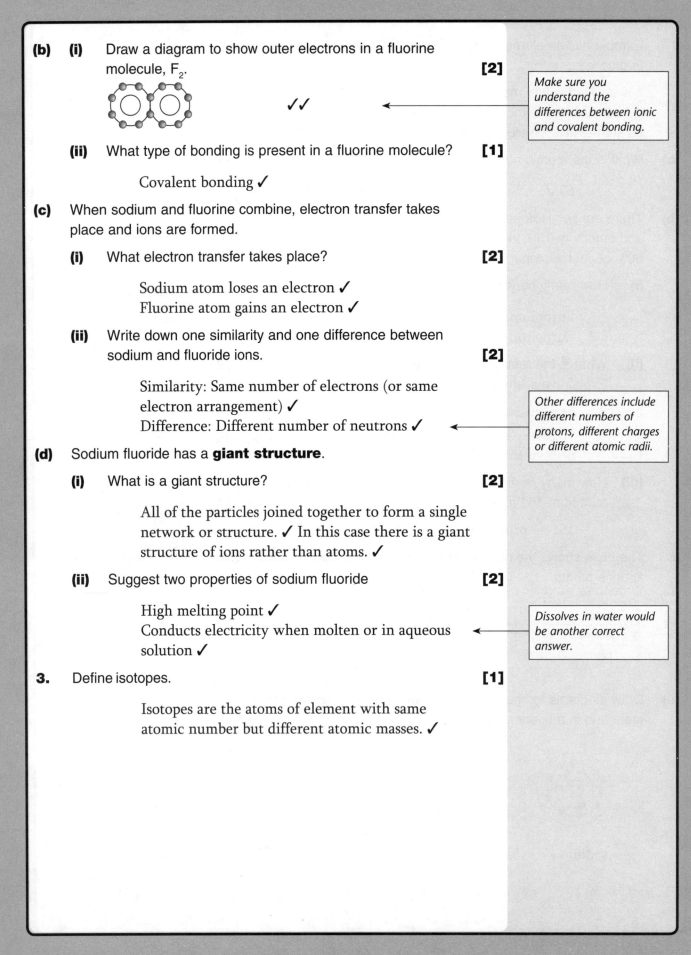

Make sure you understand the differences between ionic and covalent bonding.

(ii) What type of bonding is present in a fluorine molecule? **[1]**

Covalent bonding ✓

(c) When sodium and fluorine combine, electron transfer takes place and ions are formed.

(i) What electron transfer takes place? **[2]**

Sodium atom loses an electron ✓
Fluorine atom gains an electron ✓

(ii) Write down one similarity and one difference between sodium and fluoride ions. **[2]**

Similarity: Same number of electrons (or same electron arrangement) ✓
Difference: Different number of neutrons ✓

Other differences include different numbers of protons, different charges or different atomic radii.

(d) Sodium fluoride has a **giant structure**.

(i) What is a giant structure? **[2]**

All of the particles joined together to form a single network or structure. ✓ In this case there is a giant structure of ions rather than atoms. ✓

(ii) Suggest two properties of sodium fluoride **[2]**

High melting point ✓
Conducts electricity when molten or in aqueous solution ✓

Dissolves in water would be another correct answer.

3. Define isotopes. **[1]**

Isotopes are the atoms of element with same atomic number but different atomic masses. ✓

Exam practice questions

1. Hydrogen and chlorine react together to form hydrogen chloride.

$H_2 + Cl_2 \rightarrow 2HCl$

(a) Dry hydrogen chloride gas contains hydrogen chloride molecules.

(i) Draw a dot and cross diagram of a hydrogen chloride molecule. **[2]**

(ii) What type of bonding is present in dry hydrogen chloride molecules? **[1]**

(b) Hydrogen chloride dissolves in water to form a solution, which conducts electricity.

Explain the changes in bonding which occur when hydrogen chloride dissolves in water. **[3]**

2. The table below gives information about four substances labelled A–D.

Substance	Melting point in °C	Boiling point in °C	Electrical conductivity when solid	Electrical conductivity when molten
A	801	1470	poor	good
B	850	1487	good	good
C	–218	–183	poor	poor
D	2900	very high	poor	poor

This question is marked out of eight marks for the correct science in your answer. In addition one mark is allocated for the examiner to use to reward some aspect of Quality of Written Communication (QWC). In this case the mark will be awarded for an answer written in proper sentences with a capital letter and a full stop.

What does the data in the table show about the structures of these substances?

Explain your reasoning. **[8+1]**

3. Write the electronic configuration of the following:

(a) $^{23}_{11}Na$

(b) $^{40}_{20}Ca$ **[2]**

4. Identify isotopes from the following:

Element A: 17p, 18n, 17e

Element B: 17p, 20n, 17e

Element C: 18p, 18n, 18e **[1]**

Exam practice questions

5. Complete the following table:

Element	Proton Number (At. No.)	Nucleon Number (At. Mass)	Valency	Number of Protons	Number of Neutrons	Number of Electrons
O	*8*	*16*	*–2*		*8*	
N	*7*	*14*			*7*	*10*
F	*9*	*18*	*–1*			*10*
K	*19*	*39*	*+1*			

6. Why is the valency of oxygen –2 and not +2? Oxygen is represented as $^{16}_{8}O$. **[2]**

7. Copper has the structure of a typical metal. It has a lattice of positive ions and a 'sea' of mobile electrons. The lattice can accommodate ions of a different metal.

Give a different use of copper, that depends on each of the following:

(a) the ability of the ions in the lattice to move past each other **[1]**

(b) the presence of mobile electrons **[1]**

(c) the ability to accommodate ions of a different metal in the lattice. **[1]**

8. Fill in the blanks:

(a) Element X has _____ valency when it gains 3 electrons. **[1]**

(b) Element Y has _____ valency when it loses 1 electron. **[1]**

(c) Element Z has electronic configuration 2, 8, 8. Its valency will be _____. **[1]**

(d) The first two shells can hold _____ number of electrons. **[1]**

Organic chemistry

The following topics are covered in this section:

- ***Names of compounds and homologous series***
- ***Macromolecules***

3.1 Names of compounds and homologous series

After studying this section you should be able to:

- ***understand the concept of homologous series***
- ***explain the existence of different isomers for the same molecular formula***
- ***understand the splitting of crude oil into saleable products by fractional distillation***
- ***understand that there are different families of hydrocarbons including alkanes and alkenes***
- ***explain the different products which are formed when hydrocarbons burn in different amounts of oxygen***
- ***draw the structures of simple alcohols***
- ***describe the manufacture of ethanol by different methods***
- ***describe structural formulae of carboxylic acids and manufacture of ethanoic acid from ethanol***
- ***describe the production of esters.***

Organic compounds

- Organic compounds are those with molecules containing carbon and one or more other elements.
- There is a wide variety of organic compounds, depending on the length of the carbon chain, the elements it is attached with and their structural arrangement in space.
- All living things are made of organic compounds

Homologous series

- Homologous is derived from Greek *homos* (same) and *logos* (ratio).
- It is a series of chemical compounds having the same functional group but differing in composition by a fixed group of atoms.
- Functional groups determine most of the chemical properties of compounds.
- Compounds in a homologous group have the following points in common.

- **Similar name endings**
- **Similar chemical structure**
- **Similar chemical behaviour**
- **Can be represented by a general formula.**

Most of the simple organic compounds can be broadly classified into the following homologous groups.

'Meth' stands for one carbon atom, **'Eth'** for two, **'Prop'** for three, **'But'** for four, **'Pent'** for five and so on.

1. **Alkanes:** The names of chemicals in this group ends with 'ane' e.g. methane, ethane, propane, butane etc.

2. **Alkenes:** The names in this group end with 'ene' e.g. ethene, propene, butene, pentene etc.

3. **Alcohols:** The names in this group end with 'ol' e.g. methanol, ethanol, propanol, butanol etc.

4. **Carboxylic acids:** The names in this group end with 'oic' e.g. methanoic acid, ethanoic acid, propanoic acid, butanoic acid etc.

KEY POINT

The properties in a homologous series change with the change in number of carbon atoms per molecule.
More the number of carbon atoms present, the higher are the melting and boiling points.

'Methane' is the main constituent of natural gas.

Most of the man-made organic compounds are derived from crude oil, which along with coal and natural gas, is the most widely used fuel at present.

Isomerism

With alkanes containing up to three carbon atoms, there is only one possible structure for each molecular formula.

However, when there are four or more carbon atoms in an alkane, it is possible to have different structures.

KEY POINT

Isomerism is the existence of two or more compounds with the same molecular formula but different structural formulae.

Do not confuse the term isomer with isotope or allotrope.

The two isomers of butane are:

```
                        H   H   H
  H   H   H   H         |   |   |
  |   |   |   |     H - C - C - C - H
H-C - C - C - C - H     |   |   |
  |   |   |   |         H   |   H
  H   H   H   H           H-C-H
                            |
                            H
```

butane   2-methylpropane.

These two isomers have similar chemical properties but different physical properties.

In each structure, every carbon atom is joined to 4 other atoms and every hydrogen atom is joined to one other atom.

Isomerism becomes more common with higher alkanes, e.g. there are 75 isomers of decane $C_{10}H_{22}$. Isomers also occur with other homologous series (families), e.g. alkenes, and also with compounds in different homologous series e.g. C_2H_5OH (an alcohol) and CH_3OCH_3 (an ether).

PROGRESS CHECK

1. *For each of the following statements write true or false:*
   (a) *Isomers have the same molecular formula.*
   (b) *Isomers have the same relative formula mass.*
   (c) *Isomers have the same number of bonds.*
   (d) *Isomers have the same melting and boiling points.*
2. *Draw the three isomers of pentane,* C_5H_{12}.
3. *Butene,* C_4H_8, *is an alkene and contains a carbon–carbon double bond. Draw the structural formulae of the isomers of butene.*

1. A. True; B. True; C. True; D. False.
2. 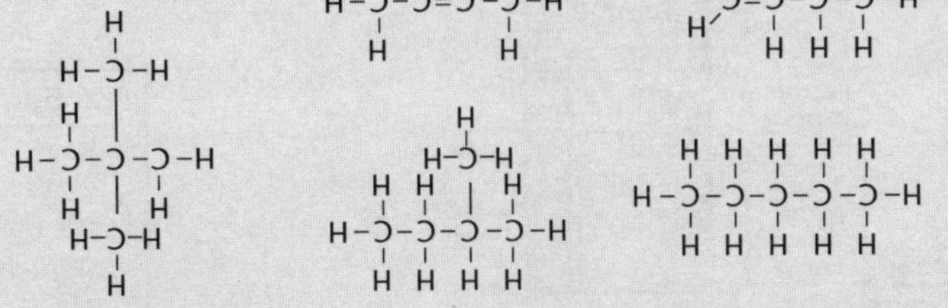
3.

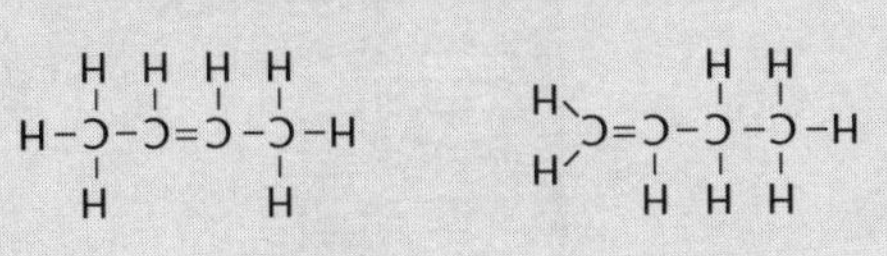

Refining crude oil

Crude oil is a **mixture** of **hydrocarbons**. Hydrocarbons are **compounds** of **carbon** and **hydrogen only**.

Most of the hydrocarbons belong to a family called **alkanes**.

Table 3.1 contains information about the first six members of the alkane family.

Remember the names of alkanes end in -ane. The prefix tells you the number of carbon atoms. Pentane contains 5 carbon atom.

Name	Formula	Structure	State at room temp	Melting point	Boiling point
Methane	CH_4	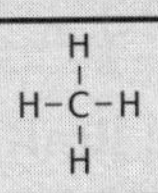 H H-C-H H	Gas	Increases down the family ↓	Increases down the family ↓
Ethane	C_2H_6	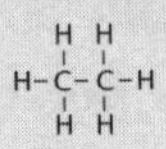 H H H-C-C-H H H	Gas		
Propane	C_3H_8	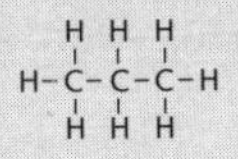 H H H H-C-C-C-H H H H	Gas		
Butane	C_4H_{10}	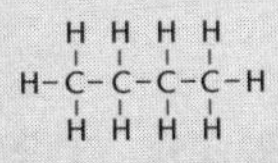 H H H H H-C-C-C-C-H H H H H	Gas		
Pentane	C_5H_{12}	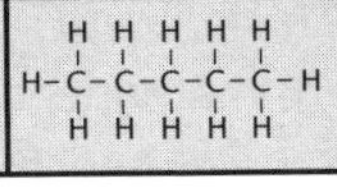 H H H H H H-C-C-C-C-C-H H H H H H	Liquid		

Alkanes

- are all **saturated hydrocarbons** (containing only single carbon–carbon bonds)
- all fit a formula C_nH_{2n+2}
- burn in air or oxygen.

Candidates often confuse saturated here with saturated when referred to solutions.

Crude oil is sometimes called petroleum.

Crude oil is separated into separate **fractions** in an **oil refinery**. This is done by **fractional distillation**. Each fraction contains hydrocarbons which boil within a **temperature range**. Each fraction has a different use.

Crude oil vapour enters a tall column and cools. The lower the boiling point, the higher the position of condensation of vapour in the column.

Fig. 3.1 shows how different fractions can be obtained from crude oil, and the different uses of these fractions are shown in the table below.

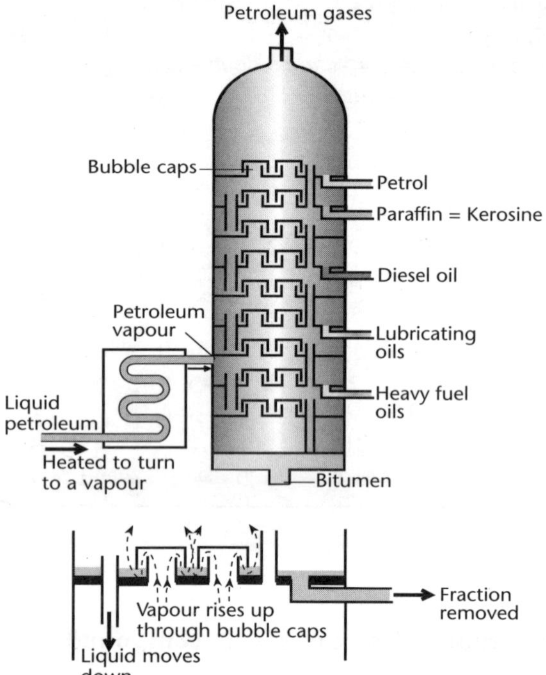

Fraction	Use
Petrol	*Fuel for cars*
Paraffin	*Fuel for aircraft*
Diesel oil	*Fuel for cars, trains*
Lubricating oil	*For motor engines*
Heavy fuel oils	*For heating*
Bitumen	*Road tar*

Fig. 3.1 Fractional distillation of crude oil

Burning alkanes

When alkanes are burned in **excess air** or oxygen, **carbon dioxide** and **water** are produced.

e.g. methane + oxygen → carbon dioxide + water

$$CH_4 + 2O_2 \rightarrow CO_2 + 2H_2O$$

Alkanes burn in a **limited supply of air** to produce **carbon monoxide** and **water vapour**. Carbon monoxide is very **poisonous**.

e.g. methane + oxygen → carbon monoxide + water

$$2CH_4 + 3O_2 \rightarrow 2CO + 4H_2O$$

Carbon monoxide has no smell. Every year in the UK up to 50 people die of carbon monoxide poisoning. Often these deaths are caused by gas appliances

Formation of free radicals

Alkanes such as methane, although largely unreactive, react with halogens such as chlorine in the presence of sunlight due to the formation of free radicals.

The reaction with chlorine atoms (free radicals) takes place, as follows:

$$Cl\,(g) + CH_4\,(g) \longrightarrow CH_3\,(g) + HCl(g)$$

(Methyl free radical)

The methyl free radical thus formed further reacts as follows:

$$CH_3(g) + Cl_2(g) \longrightarrow CH_3Cl(g) + Cl(g)$$

(Chlorine free radical)

The chlorine free radical further reacts with methane and a **chain reaction** follows. The final result of this reaction is:

$$\underset{\text{(Methane)}}{CH_4(g)} + \underset{\text{(Chlorine)}}{Cl_2(g)} \xrightarrow{\text{Sunlight}} \underset{\text{(Chloromethane)}}{CH_3Cl(g)} + \underset{\text{(Hygrogen chloride)}}{HCl(g)}$$

In this reaction, hydrogen atom of methane gets substituted by a chlorine atom and the reaction is therfore called a **substitution reaction.**

Alkenes

- Alkenes are a series of homologous organic compounds, with the general formula $\mathbf{C_nH_{2n}}$.
- They are **hydrocarbons** like alkanes, but are more reactive than them.
- They possess a double covalent i.e. **unsaturated** bond.

One of the two bonds in double covalent bonds can be broken to add extra atoms in the molecule i.e. the bond is not fully saturated and there is scope for other atoms to be added, hence it is called **unsaturated**.

Table 3.2 shows the molecular structure of the first four alkenes:

Table 3.2

Name	Formula	Molecular Structure
Ethene	C_2H_4	H H C = C H H
Propene	C_3H_6	H H H H—C —C = C H H
Butene	C_4H_8	H H H H H—C— C —C = C H H H
Pentene	C_5H_{10}	H H H H H H—C —C— C —C = C H H H H

Alcohols

The structural formulae of the first four members of the alcohol family are given in **Table 3.3**.

Name	Formula	Molecular structure
Methanol	CH_3OH	H H H–C–O–H H
Ethanol	CH_3CH_2OH	H H H–C–C–O–H H H
Propanal	$CH_3CH_2CH_2OH$	H H H H–C–C–C–O–H H H H
Butanol	$CH_3CH_2CH_2CH_2OH$	H H H H H–C–C–C–C–O–H H H H H

The functional group in alcohols is –OH.

Ethanol C_2H_5OH is an organic chemical of great importance. It is a member of the **alcohol** family. Alcohols have a general formula $C_nH_{2n+1}OH$.

Alcohols are derived from an alkane by removing an –H and adding an –OH.

The structural formula of ethanol is

```
   H  H
   |  |
H- C- C-O-H
   |  |
   H  H
```

Manufacture of ethanol

Large quantities of ethanol are manufactured for industrial use.

There are two methods of producing ethanol – from **ethene** or from **sugar**.

From ethene

Large amounts of ethene are produced from **cracking fractions** from **crude oil**.

Much of this is used to make poly(ethene) and ethanol. Formation of ethanol involves an **addition reaction**.

Ethene is mixed with steam and passed over a phosphoric acid catalyst at 600°C and at high pressure.

$$C_2H_4(g) + H_2O(g) \rightarrow C_2H_5OH(g)$$

ethene + steam → ethanol

$$\begin{matrix} H & & H \\ & C{=}C & \\ H & & H \end{matrix} \;+\; \begin{matrix} H & & H \\ & O & \end{matrix} \;\rightarrow\; \begin{matrix} & H & H & \\ H- & C- & C- & O-H \\ & H & H & \end{matrix}$$

From sugar

A dilute solution of ethanol is present in beer and wine.

KEY POINT
- **Ethanol can be prepared by the fermentation of sugar solutions using enzymes in yeast.**

The solution is kept in a warm place for several days.

It is actually a dilute solution of ethanol.

Spirits such as whisky and gin are produced by fractional distillation.

$$C_6H_{12}O_6(aq) \rightarrow 2C_2H_5OH(aq) + 2CO_2(g)$$

glucose → ethanol + carbon dioxide

A more concentrated solution of ethanol is produced by **fractional distillation**.

There are good opportunities here to compare the manufacture of ethanol by two different methods. There is no better method. It depends upon circumstances.

The method used to manufacture ethanol depends upon the materials available.

1. In developed countries, such as the United States and in Europe, there are large amounts of ethene available. Making ethanol from ethene would be preferred.
2. In countries such as Mauritius, which do not have crude oil but do have sugar produced from sugar cane in large amounts, fermentation would be preferred.

Ethanol made by the fermentation process is a **batch** process. Ethanol is produced from ethene by a **continuous** process.

Table 3.4 below compares the advantages and disadvantages of each method.

Ethanol by fermentation	Ethanol from ethene
Advantages	**Advantages**
Uses renewable resources e.g. sugar cane Uses waste materials	Fast reaction, Continuous process Does not need large reaction vessels Produces pure ethanol
Disadvantages	**Disadvantages**
Large volume needed to produce small amount of ethanol Needs large reaction vessels Fractional distillation is expensive Fermentation is slow When ethanol reaches a certain concentration, the reaction stops	Uses a non-renewable resource Energy is needed to produce steam High percentage of ethene remains unreacted and must be recycled

Uses of ethanol

When perfume is sprayed on the skin, the cooling felt is due to the evaporation of ethanol.

1. Ethanol is widely used as a **solvent**. It is also used in paints, varnishes, perfumes etc.
2. Ethanol is used as a **fuel**. In Brazil, either pure ethanol or a mixture of petrol and ethanol is used as a fuel in cars. Ethanol burns in excess air to form carbon dioxide and water.

   $C_2H_5OH + 3O_2 \rightarrow 2CO_2 + 3H_2O$

3. Ethanol is used for **making other organic chemicals** e.g. ethanoic acid, esters.
4. Ethanol is used in **alcoholic drinks**.

Different drinks contain different percentages of alcohol (**see Table 3.5** below)

Drink	Approximate percentage of ethanol
Beers	4
Wine	12
Fortified wine e.g. sherry	18
Spirits (whisky)	35

However, there are some harmful effects of ethanol including:

- impaired coordination and judgement
- slower reaction times
- promotes aggression
- causes depression and other mental disorders
- causes ulcers, high blood pressure, brain and liver damage.

Pure ethanol cannot be purchased in shops.

We can buy **methylated spirits**. This is ethanol with added **methanol**. Methanol is highly toxic. Other substances are added to make it undrinkable and a purple dye is added as a warning.

PROGRESS CHECK

1. *Write the molecular formula of methanol.*
2. *Draw the structural formula of methanol.*
3. *Write an equation for the combustion of methanol in excess air.*
4. *Draw the structural formulae of two alcohols with a molecular formula C_3H_8O.*
5. *Ethene, C_2H_4, is produced when ethanol vapour is passed over heated aluminium oxide. Write the equation for this reaction.*
6. *What type of reaction is taking place?*

1. CH_4O or CH_3OH
2. H–C(H)(H)–O–H
3. $2CH_3OH + 3O_2 \rightarrow 2CO_2 + 4H_2O$;
4. H–C(H)(H)–C(H)(H)–C(H)(H)–O–H   H–C(H)(H)–C(H)(OH)–C(H)(H)–H
5. $C_2H_5OH \rightarrow C_2H_4 + H_2O$;
6. Removal of water (dehydration).

Carboxylic acids

Carboxylic acids form a series of homologous organic compounds with the general formula $\mathbf{C_nH_{2n+1}COOH}$.

They are **weak acids,** with the functional group — **COOH**.

Table 3.6 Formula and chemical structure of some alcohols

Name	Formula	Molecular structure
Methanoic acid	*HCOOH*	H—C(=O)—OH
Ethanoic acid	CH_3COOH	CH_3—C(=O)—OH
Propanoic acid	CH_3CH_2COOH	CH_3—CH_2—C(=O)—OH
Butanoic acid	$CH_3CH_2CH_2COOH$	CH_3—CH_2—CH_2—C(=O)—OH

The functional group in carboxylic acid is- COOH

Ethanoic acid, which is the main constituent of vinegar, is also the most well known carboxylic acid.

Ethanoic acid

Ethanoic acid is a weak acid with a molecular formula $C_2H_4O_2$.

It has a structural formula given below.

$$H_3C-C(=O)-O-H$$

Ethanoic acid used to be called acetic acid. Vinegar is a dilute solution of ethanoic acid.

It is called ethanoic acid because, like ethane, it contains two carbon atoms.

Ethanoic acid is prepared in industry by passing ethanol and air over a heated catalyst.

Wine containing ethanol goes sour when left in contact with air. This is due to bacterial oxidation.

$$C_2H_5OH + O_2 \rightarrow CH_3COOH + H_2O$$

Properties of ethanoic acid

Ethanoic acid is a weak acid. It has similar reactions to other acids.

1. Indicators

Ethanoic acid turns litmus paper red. Solutions of ethanoic acid have a pH value of about 4.

2. Metals

Ethanoic acid reacts with reactive metals, e.g. magnesium to produce a salt plus **hydrogen**.

$Mg + 2CH_3COOH \rightarrow (CH_3COO)_2Mg + H_2$

magnesium + ethanoic acid → magnesium ethanoate + hydrogen

3. Metal oxides (bases)

Ethanoic acid reacts with a base to form a **salt** and **water** only.

$CuO + 2CH_3COOH \rightarrow (CH_3COO)_2Cu + H_2O$

copper(II) oxide + ethanoic acid → copper(II) ethanoate + water

4. Metal carbonates

Ethanoic acid reacts with a carbonate to produce salt and water and **carbon dioxide**.

$Na_2CO_3 + 2CH_3COOH \rightarrow 2CH_3COONa + H_2O + CO_2$

Sodium carbonate + ethanoic acid → sodium ethanoate + water + carbon dioxide
(salt)

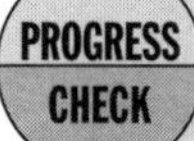

Methanoic acid, HCOOH, is a weak acid.

1. Draw the structural formula of methanoic acid.

Questions 2–8. Reagents are added to methanoic acid. Fill in the gaps in the table.

Test	Result of test	Name of substances formed
Litmus	*2*	
Add magnesium	*3*	*4*
Add copper(II) oxide	*5*	*6*
Add sodium carbonate	*7*	*8*

1. H–C(=O)–O–H

2. Turns red; 3. Bubbles of colourless gas; 4. Hydrogen and magnesium methanoate; 5. Blue solution; 6. Copper(II) methanoate and water; 7. Bubbles of colourless gas; 8. Sodium methanoate, carbon dioxide and water.

Esters

KEY POINT **Esters are compounds formed by the reaction of an organic acid with an alcohol.**

Ethyl ethanoate is an ester produced by reaction of an acid (**ethanoic acid**) and an alcohol (**ethanol**).

$$CH_3COOH + C_2H_5OH \rightleftharpoons CH_3COOC_2H_5 + H_2O$$

ethanoic acid + ethanol ⇌ ethyl ethanoate + water

You will remember that acid + alkali → salt + water. Now remember acid + alcohol ⇌ ester + water.

A mixture of ethanol and ethanoic acid are heated under reflux (**Fig. 3.2** below) with a little concentrated sulphuric acid.

Heating under reflux prevents the reactants being lost by evaporation. If the mixture boils, the vapour is condensed again and drops back into the flask.

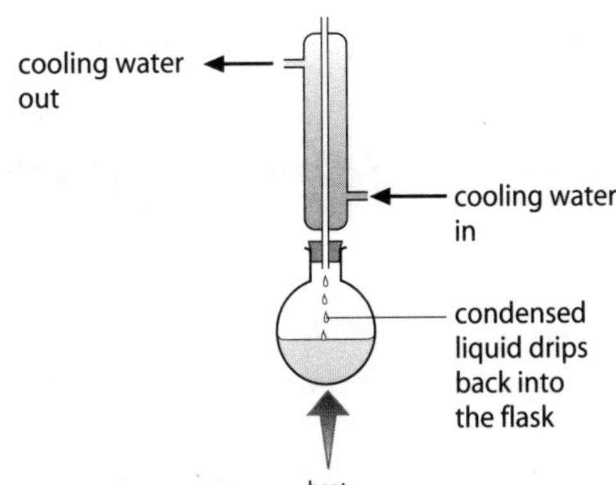

Concentrated sulphuric acid removes the water and pushes the reaction to the right.

Fig. 3.2

The structural formula of ethyl ethanoate is

```
   H     O
   |   //      H  H
H- C - C       |  |
   |    \O  -  C- C-H
   H           |  |
               H  H
```

Terylene, a synthetic polymer, is also like fats, made up of molecules joined by ester link (OCO).

Esters are **sweet-smelling liquids**. They are used as **flavouring agents**.

Ethyl ethanoate smells like pears and is used for flavouring sweets.

Natural fats and oils are naturally occurring esters derived from long chain acids e.g. $CH_3(CH_2)_{16}COOH$ and propane-1,2,3-triol (glycerol) $CH_2OHCHOHCH_2OH$.

The structural formula of a natural ester is

$$\begin{array}{l} CH_2OOC(CH_2)_{16}CH_3 \\ | \\ CHOOC(CH_2)_{16}CH_3 \\ | \\ CH_2OOC(CH_2)_{16}CH_3 \end{array}$$

Fats and oils are used in the manufacture of soaps and margarine.

PROGRESS CHECK

Methyl propanoate, $C_2H_5COOCH_3$ is an ester.

1. Draw the structural formula of methyl propanoate.
2. Write down the name and formula of the acid and the alcohol that could be used to make methyl propanoate.
3. Is methyl propanoate a solid, liquid or gas at room temperature?
4. Another ester is an isomer of methyl propanoate. Write the name and draw the structure of the isomer.
5. An acid has the same molecular formula as methyl propanoate. Draw the structure of this acid.

1\.
2\. Methanol, CH_3OH, and propanoic acid, C_2H_5COOH;
3\. Liquid; 4. Ethyl ethanoate.
5\.

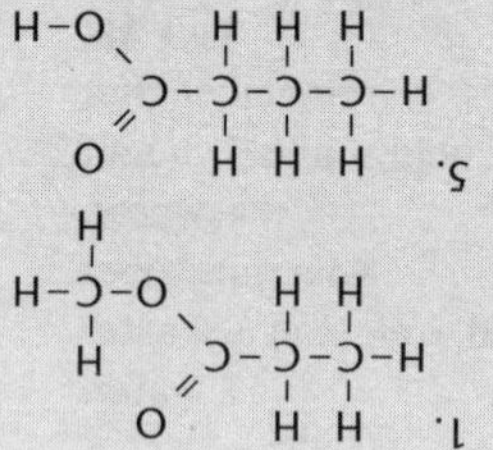

3.2 Macromolecules

LEARNING SUMMARY

After studying this section you should be able to:

- ***understand that large chain hydrocarbons can be broken into simpler molecules by cracking***
- ***explain how small alkenes can be linked together to form addition polymers***
- ***evaluate the benefits of addition polymers for a range of uses***
- ***recall the hardening of natural fats and oils by reaction with hydrogen***
- ***enumerate some condensation polymers***
- ***describe the structure of carbohydrates, proteins and fats***
- ***explain the hydrolysis of carbohydrates and proteins***
- ***recall that certain drugs are able to relieve pain.***

Making addition polymers

Higher boiling point fractions are more **difficult to sell** as there is less demand for them.

The petrochemical industry breaks up these long chains to produce short molecules. This decomposition reaction is called **cracking**.

Cracking is catalytic decomposition.

Cracking involves passing the vapour of the high boiling point fraction over a catalyst at high pressure.

Compounds such as **ethene** are produced.

Ethene

- belongs to a family of **alkenes**, C_nH_{2n}
- is an **unsaturated hydrocarbon** with a formula C_2H_4
- is a gas at room temperature
- molecules contain a **carbon–carbon double bond**.

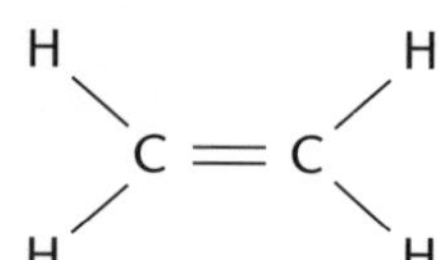

Fig. 3.3 Ethene

A common mistake here is to write that the solution turns clear. This is incorrect. All solutions are clear.

There is a simple test to distinguish ethane and ethene. If ethene gas is bubbled through a solution of **bromine,** the solution changes from **red–brown** to **colourless**.

ethene + bromine → 1,2-dibromoethane

$C_2H_4 + Br_2 \rightarrow C_2H_4Br_2$

In an addition reaction two molecules combine to form a single product.

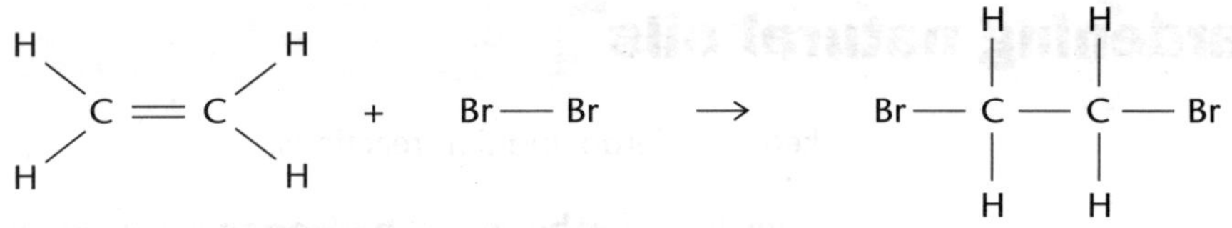

Fig. 3.4 Addition of bromine to ethene

This is an example of an **addition reaction**. Two reactants react to form a single product and the double bond in ethene becomes a single bond.

There is no colour change when ethane is added to a solution of bromine.

Small ethene molecules, produced by cracking, are joined together by a process called addition polymerisation.

Ethene is called the **monomer** and **poly(ethene)** is called the **addition polymer**. In order to produce this polymer, the ethene vapour is passed over a heated catalyst. A series of addition reactions occur.

Notice that the monomer contains a double bond and this becomes a single bond when the molecules join together. The chains can have thousands of units added together. The properties of a sample of polymer depend upon the length of the chain.

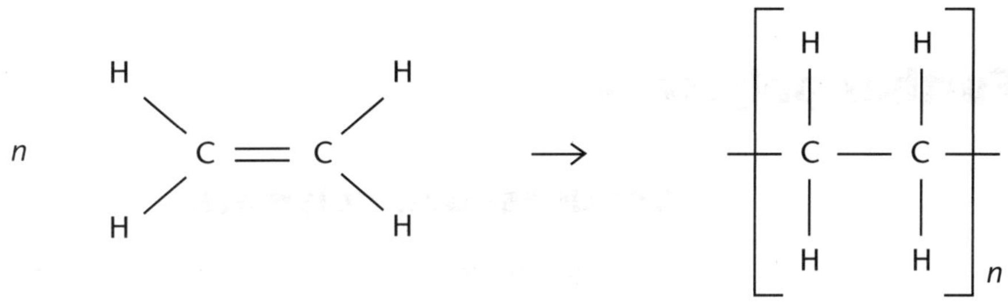

Fig. 3.5 Polymerisation of ethene

Uses of addition polymers

Addition polymers such as poly(ethene) and poly(vinyl chloride) have many uses. They have replaced traditional materials such as metals, paper, cardboard and rubber.

Common uses include:

poly(ethene) – wrappings for food, storage containers, milk crates

poly(vinyl chloride) – wellington boots, insulation for electrical wiring.

Table 3.7 compares some of the advantages and disadvantages of polymers.

Advantages of polymers	Disadvantages of polymers
Do not absorb water	*Do not rot away and can cause litter problems*
Can be moulded into shape	*Not easy to recycle as there are many types*
Can be coloured	*Burn to form poisonous fumes*
Low density	
Strong	

KEY POINT

Alkanes are saturated hydrocarbons used as fuels and chemical feedstock. Alkanes can be cracked to produce alkenes and alkenes can be polymerised to form addition polymers.

Hardening natural oils

Alkenes undergo addition reactions.

If a mixture of **ethene** and **hydrogen** is passed over a heated nickel catalyst, an addition reaction can take place, the product of which is **ethane**.

$$H_2C=CH_2 + H-H \rightarrow H_3C-CH_3$$

Natural oils such as sunflower oil are **liquid** and **unsaturated** i.e. they contain carbon–carbon double bonds.

These oils can be **hardened** by addition reactions with hydrogen. The oil and hydrogen are passed over a nickel catalyst at 170°C.

The resulting fat is used as margarine.

Further polymers

Condensation polymers

Some general characteristics of this class of compounds are given below:

- **are more complex than addition polymers**
- **are formed by reaction of two different chemicals**
- **a water molecule is lost during each reaction and that is why the process is called condensation polymerisation**
- **some common types are Nylon, Terylene(Polyester) etc.**

Nylon

It is formed by the following reaction:

(1,6-Diaminohexane) + (Hexanedioic acid) → Nylon + Water

$$H_2N(CH_2)_6NH_2 + HOOC(CH_2)_4COOH \longrightarrow H_2N(CH_2)_6NHOC(CH_2)_4COOH + H_2O$$

Amide Link (arrow pointing to NHOC)

The structure consists of a series of two different molecules joined together by an **amide** link.

Amide link is also found in *proteins.*

The structure of polyamides, like nylon, is represented by the following chain :

Nylon is also called polyamide since an amide link is formed during its polymerisation

```
  O        O                O                 O
  ||       ||               ||                ||
 –C–■–C–N–□–C–N–■–C–N–□–N–
            |               |                 |        |
            H               H                 H        H
```

Nylon is used to produce yarns which are used in the manufacture of cloths, ropes, racket strings etc.

Terylene

It is formed by the following reaction :

(ethane-1,2 diol) + (benezene-1,4-dicarboxylic acid) ⟶ Terylene + Water

$HO(CH_2)_2OH + HOOC(C_6H_4)COOH \longrightarrow HON(CH_2)_2OCO(C_6H_4)COOH + H_2O$

Ester Link

Ester link is also found in *fats.*

The structure consists of a series of two different molecules joined together by an **ester** link.

The structure of polyesters, like terylene, is represented by the following chain :

Terylene is also called *polyester* since an ester link is formed during its polymerisation

```
O       O             O              O
||      ||            ||             ||
C–■–C–O–□–C–O–■–C–O–□–O–
```

Terylene, like nylon, is used to produce yarns which are also used in the manufacture of various fabrics etc.

Thermoplastic and thermosetting polymers

Polymers can be classified as **thermoplastic** or **thermosetting**.

Thermoplastic polymers e.g. poly(ethene) melt easily when heated. Thermosetting polymers e.g. Bakelite do not melt when heated. On stronger heating they decompose.

There is a difference in structure between thermoplastic and thermosetting polymers.

A simple representation of the two types of polymer is shown in **Fig. 3.6** below. In a thermoplastic polymer the chains are not linked. On melting, the chains are able to move freely over each other. In a thermosetting polymer, there are strong links between the polymer chains. The rigid structure is not easily broken down.

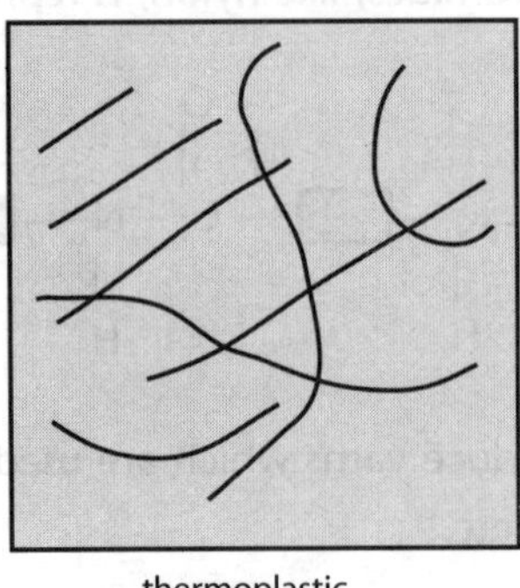

thermoplastic

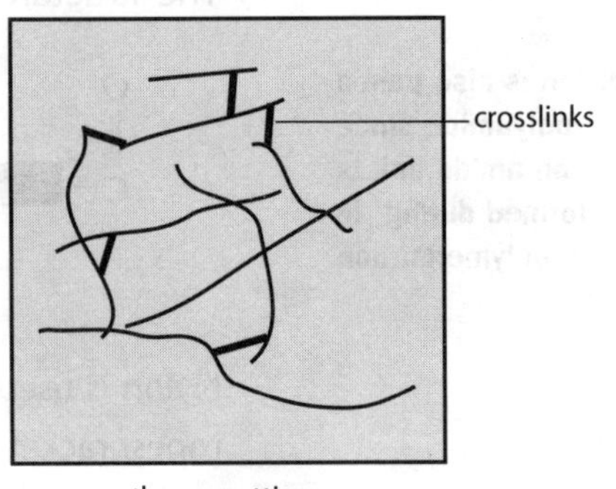

thermosetting

Fig. 3.6

Natural rubber consists of chains of polymer molecules. It is naturally soft and sticky. It can be hardened by a process of **vulcanisation** where sulphur atoms link the chains by cross-linking.

Disposal of polymers

Carbon dioxide produced increases the greenhouse effect.

Unlike materials such as paper, cardboard and wood, polymers do not rot away when tipped in landfill sites. They are said to be **non-biodegradable**.

Recycling polymers is not economic as the costs of collection and sorting are greater than the costs of making new polymer.

Polymers can be **incinerated**. **Carbon dioxide** and **water vapour** are produced.

KEY POINT

Some polymers can produce toxic gases: hydrogen cyanide from polymers containing nitrogen, hydrogen chloride from polymers containing chlorine.

PROGRESS CHECK

1. *A polymer melts, when heated, without decomposing. Is this polymer thermoplastic or thermosetting?*
2. *Which type of polymer, thermosetting or thermoplastic, is most suitable for*
   - (a) *electric light fittings which must withstand heat;*
   - (b) *plastic sheet which may be moulded to make trays;*
   - (c) *heat-resistant saucepan handles;*
   - (d) *plastic soldiers?*

Three polymers are

poly(ethene) **poly(acrylonitrile)** **poly(chloroethene)**

3. *Which polymer could produce hydrogen chloride on combustion?*
4. *Which polymer could produce hydrogen cyanide?*
5. *Why is it more difficult to recycle thermosetting polymers than thermoplastic polymers?*

1. Thermoplastic; 2(a). Thermosetting; (b) Thermoplastic; (c) Thermosetting; (d) Thermoplastic; 3. Poly(chloroethene); 4. Poly(acrylonitrile); 5. They cannot be melted and remoulded without decomposition.

Carbohydrates, proteins and fats

KEY POINT **Carbohydrates are compounds containing carbon, hydrogen and oxygen fitting a general formula $C_x(H_2O)_y$.**

A balanced diet contains carbohydrates, proteins and fats. There are also smaller quantities of vitamins, minerals, etc. Other chemicals such as food additives and raising agents may be added.

Carbohydrates may be divided into **monosaccharides** (e.g. glucose and fructose, both $C_6H_{12}O_6$), **disaccharides** (e.g. sucrose and maltose, both $C_{12}H_{22}O_{11}$) and **polysaccharides** (e.g. starch and cellulose). Polysaccharides are polymers of monosaccharides.

Fig. 3.7 below shows two structures – a chain structure and a ring structure. The ring structure better explains the structure of glucose.

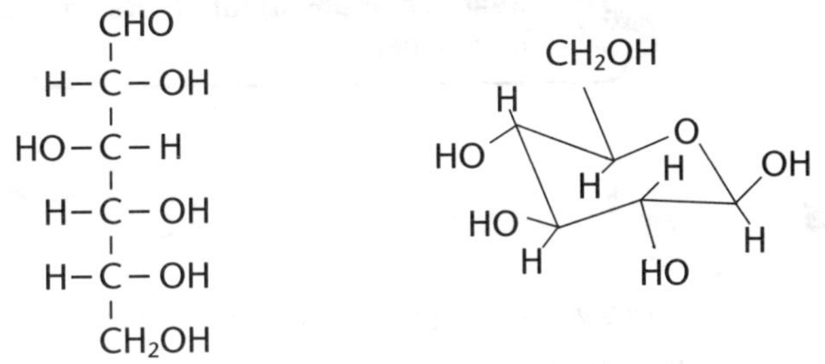

```
    CHO
     |
  H—C—OH
     |
 HO—C—H
     |
  H—C—OH
     |
  H—C—OH
     |
    CH2OH
```

Fig. 3.7

Sucrose is the carbohydrate we commonly call sugar. A reaction where two molecules are joined and a small molecule lost is called a condensation reaction.

Sucrose is formed when a **glucose** and a **fructose** molecule join together with the loss of a water molecule.

Fig. 3.8 shows a simple representation of a sucrose molecule.

Starch and cellulose are **condensation polymers (Fig. 3.9).**

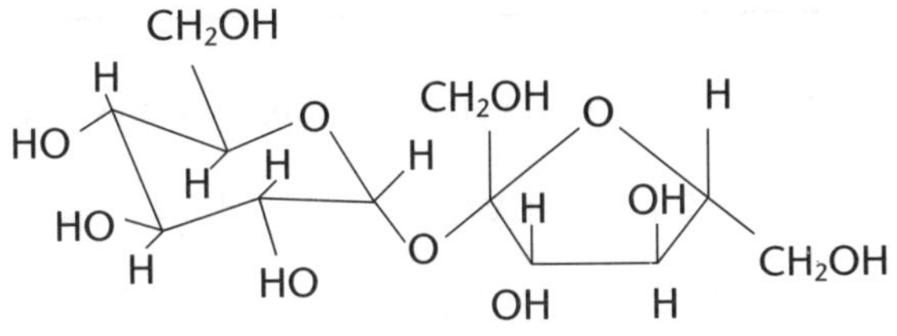

Fig. 3.8 Sucrose

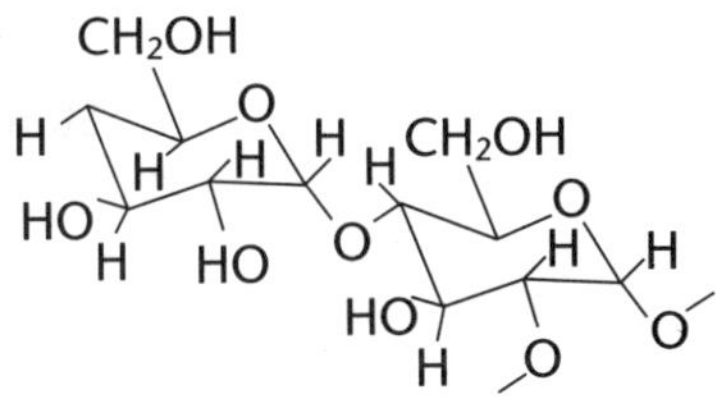

Fig. 3.9 Starch

KEY POINT **Proteins are condensation polymers made by linking together a large number of amino acid molecules.**

Fig. 3.10 shows how two amino acid molecules are linked together by a **peptide link** to form a **dipeptide**.

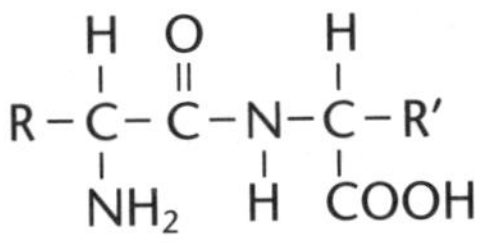

Fig. 3.10

Peptide link is also found in nylon, a synthetic polymer (as discussed previously). It is also called **amide** link.

KEY POINT **A protein is made when many amino acid molecules are joined together by peptide links.**

A single protein can contain as many as 500 amino acid units combined together. A protein can be pictured as in **Fig. 3.11** below as coils, with the loops of the coils held in position by weak crosslinks.

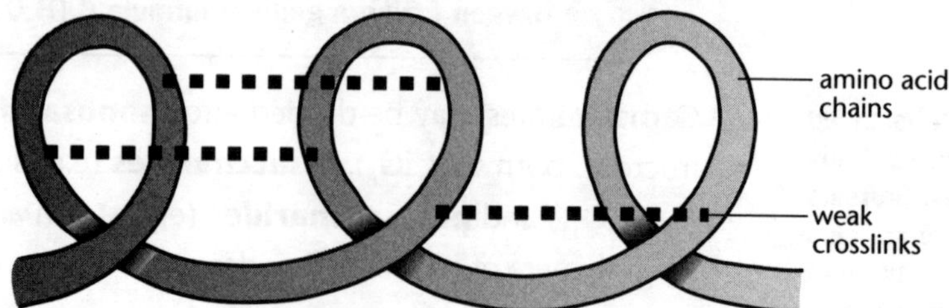

Fig. 3.11

Fats are esters. Hydrolysis of fats produces soaps and glycerol. Reaction of unsaturated fats and oils with hydrogen produces margarine.

Hydrolysis

Hydrolysis is a chemical reaction which involves the reaction of a compound with water.

Hydrolysis of carbohydrates

Hydrolysis enables carbohydrates, e.g. starch, to be used as an energy source by yielding sugars such as glucose, maltose, sucrose etc.

Starch, the most common carbohydrate, hydrolyses as follows :

$$\text{Starch} + \text{Water} \xrightarrow[\text{Heat}]{\text{Dilute HCl}} \text{Glucose}$$

$$C_6H_{10}O_5(s) + H_2O(l) \longrightarrow C_6H_{12}O_6(aq)$$

This type of hydrolysis, which takes place in presence of acids, is called **acid hydrolysis**.

In the human mouth, the reaction conditions are different and starch hydrolyses as follows:

Amylase is the enzyme present in saliva.

$$\text{Starch} + \text{Water(In Saliva)} \xrightarrow{\text{Amylase(catalyst)}} \text{Maltose}$$

$$2(C_6H_{10}O_5)(s) + H_2O(l) \longrightarrow C_{12}H_{22}O_{11}(aq)$$

This type of hydrolysis, which takes place in presence of enzymes, is called **enzyme hydrolysis**.

Hydrolysis of proteins

Hydrolysis of proteins releases the amino acids present in them. It can be used to find out which amino acids are present in a particular protein.

Proteins are hydrolysed by heating them with dilute hydrochloric acid (acid hydrolysis).The constituent amino acids, which are released in the solution, are then seperated from the mixture and identified by thin layer **chromatography**.

Chromatography

Chromatography is a technique of seperating two or more soluble solids from a mixture. It is mostly used for the identification of the constituents of a mixture.

Chromatography is used to seperate and identify the products of hydrolysis of carbohydrates and proteins.

Simplest kind is **paper chromatography**. In this technique, the liquid mixture is put on a chromatography paper and this paper is partially dipped in a suitable solvent. As the solvent is absorbed by the paper, it moves up carrying along the solids, which are then deposited at different heights on the paper depending on their solubility in the solvent.

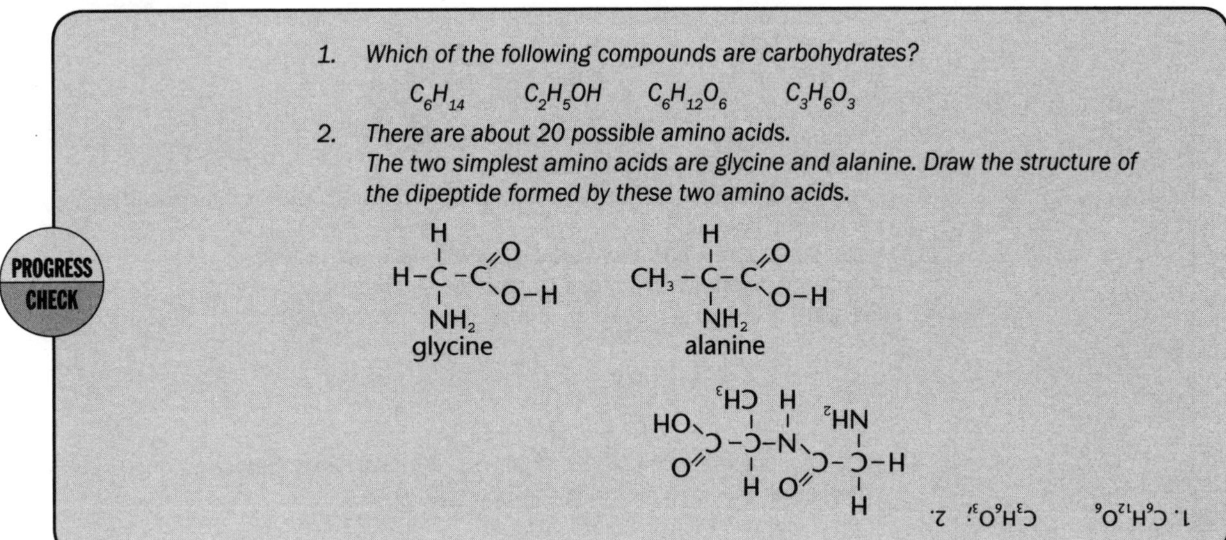

PROGRESS CHECK

1. *Which of the following compounds are carbohydrates?*

   C_6H_{14} C_2H_5OH $C_6H_{12}O_6$ $C_3H_6O_3$

2. *There are about 20 possible amino acids.*
   The two simplest amino acids are glycine and alanine. Draw the structure of the dipeptide formed by these two amino acids.

1. $C_6H_{12}O_6$ $C_3H_6O_3$; 2.

Drugs

A drug is a substance which when taken modifies or affects the chemical reactions in the body.

One group of drugs is **analgesics**. These are drugs that reduce pain. They include **aspirin**, **paracetamol** and **ibuprofen**.

KEY POINT

Aspirin is a chemical called acetyl salicylic acid. It is made in industry from salicylic acid.

salicylic acid + CH_3COCl (ethanoyl chloride) → acetyl salicylic acid (aspirin) + HCl

Fig. 3.12

Salicylic acid is found in the bark of the willow tree. Ancient Greeks and native Indians used the bark of willow trees to counter fever and pain. Salicylic acid is bitter and irritates the stomach. In 1890 *Felix Hoffman* showed the conversion of salicylic acid into acetyl salicylic acid, which makes it more suitable for intended purpose.

Fig. 3.13 below shows the structure of the sodium salt of aspirin. This is more

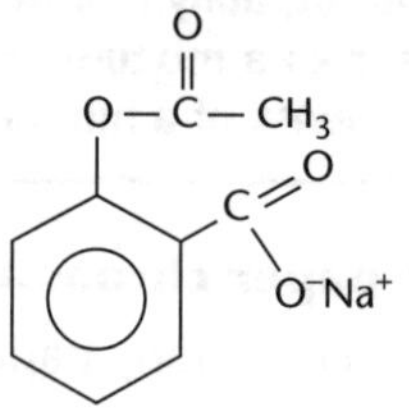

sodium salt of aspirin
(more soluble)

Fig. 3.13

PROGRESS CHECK

1. *Which one of the following compounds is not a hydrocarbon?*
   C_2H_4 C_2H_6O C_6H_6 C_4H_{10}
2. *Which one of the hydrocarbons in the list is not an alkane?*
   C_6H_{12} C_7H_{16} $C_{10}H_{22}$ $C_{40}H_{82}$
3. *Write down a use for each of the following alkanes:*
   (a) methane; (b) propane; (c) octane.
4. *LPG is used as a fuel in cars as an alternative to petrol. How is it produced in the refining process?*
5. *A colourless gas* **X** *has a formula C_3H_6. It decolourises bromine.*
   X *could be* **A.** *ethane* **B.** *ethene* **C.** *propane* **D.** *propene*
6. *Poly(vinyl chloride) is made from a monomer called vinyl chloride:*

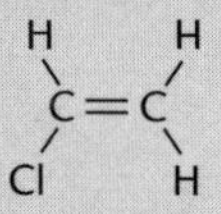

   $$\begin{matrix} H & & H \\ & C{=}C & \\ Cl & & H \end{matrix}$$

   Which, among the following, is the correct chemical name for vinyl chloride?
   chloroethane **chloroethene** **ethene**
7. *Draw the structure of poly(vinyl chloride).*
8. *Three fractions from the crude oil refinery are: kerosene, petrol and bitumen. Put these three fractions in the correct order, working down from the top of the column.*
9. *Suggest uses for each of the fractions in 8.*
10. *Decane, $C_{10}H_{22}$ can be cracked into a mixture of ethene and ethane. Write a balanced equation for this reaction.*

1. C_2H_6O; 2. C_6H_{12}; 3. (a) Household gas supply (natural gas) (b) Camping gas (c) Petrol; 4. Gases leaving the top of the column; 5. D; 6. Chloroethene ;

7. $\left[-\underset{H}{\overset{H}{C}}-\underset{Cl}{\overset{H}{C}}- \right]_n$

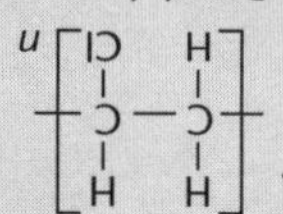

8. Petrol, kerosene, bitumen; 9. Fuel in cars; fuel in aeroplanes; road tar

10. $C_{10}H_{22} \rightarrow 4C_2H_4 + C_2H_6$

Sample IGCSE questions

1. This question is about two families of hydrocarbons – alkanes and alkenes.

Ethane, C_2H_6, is an alkane and ethene, C_2H_4, is an alkene.

Ethane: H₃C–CH₃ drawn with all bonds (H–C–C–H, each C bonded to two further H)

Ethene: H₂C=CH₂

The question gives you the structures of ethane and ethene. These are very useful to help you to answer this question.

(a) Ethane is said to be a saturated hydrocarbon and ethene is an unsaturated hydrocarbon. Explain the meaning of the terms 'saturated' and 'unsaturated'. **[2]**

A saturated hydrocarbon contains only single carbon–carbon bond ✓. An unsaturated hydrocarbon has one or more double or triple bond between carbon atoms ✓.

Your answer must explain the difference between saturated and unsaturated hydrocarbons.

(b) Ethene undergoes polymerisation reactions to produce an addition polymer called poly(ethene). **[3]**

(i) Explain what is meant by the term 'addition polymer'.

A substance formed when unsaturated monomer units ✓ join together to form a very long molecule (called a polymer) ✓ with no other substance being formed ✓.

There are three marks, so you must include three points in your answer.

(ii) Draw the structure of poly(ethene). **[2]**

$$\left(-\overset{\displaystyle H}{\underset{\displaystyle H}{C}} - \overset{\displaystyle H}{\underset{\displaystyle H}{C}} - \right)_n$$

✓✓

You must show poly(ethene) is a long chain and the double bond becomes a single bond.

(iii) Poly(ethene) is used for packaging materials. Before poly(ethene), paper or cardboard would have been used. Suggest advantages of poly(ethene) over paper and cardboard. **[3]**

Poly(ethene) does not absorb water or lose strength when wet ✓. It is easily coloured ✓ and is transparent so the contents can be seen ✓.

The question is only about advantages. Make sure the things you give are advantages. Do not give cheaper as an answer. You have no knowledge about the relative costs.

Sample IGCSE questions

2. Draw the structural formula of ethanol, C_2H_5OH. **[1]**

```
    H   H
    |   |
H - C - C - O - H
    |   |
    H   H
```
✓

3. The oxidation of ethanol can take place by reaction with an acidic solution of potassium dichromate(VI):

ethanol + orange dichromate(VI) → green chromium(III) + ethanoic acid

This is an alternative way of oxidising ethanol.

(a) Write down the names of three elements in potassium dichromate(VI) **[2]**

potassium, chromium and oxygen ✓✓

The names of the elements are given in the periodic table.

The above reaction is used in one type of breathalyser (a device revealing a person's blood alcohol content).

There is a legal limit marked on the breathalyser.

The diagram shows the breathalyser

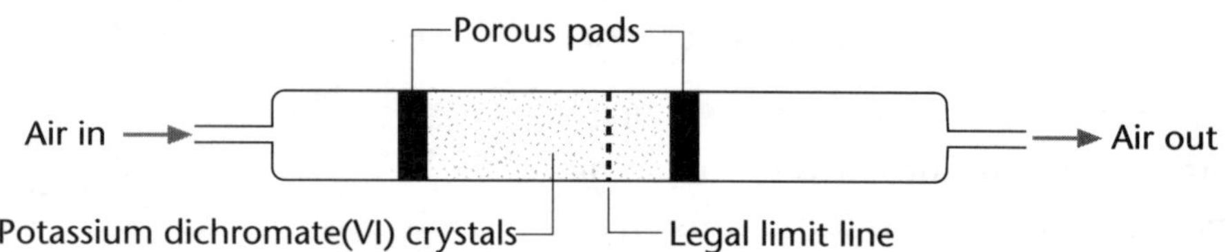

(b) How is an 'overdrunk' person, indicated by a breathanalyser? **[2]**

Orange crystals turn green ✓. If a person is heavily drunk, green colour goes beyond the legal limit mark ✓.

4. State two effects of ethanol on the human body. **[2]**

Slower reaction times ✓. Causes depression ✓.

Many students write about being drunk but their answers are not sufficiently scientific.

Exam practice questions

1. Ethanol, C_2H_5OH, can be produced from ethene, C_2H_4, or from sugar.

Some ethanol is oxidised to produce ethanoic acid.

(a) Sugar is a carbohydrate.

(i) Write down the names of the three elements present in sugar. **[3]**

(ii) What is the name of the process used to turn sugar into ethanol? **[1]**

(iii) What must be present for the change from sugar to ethanol to take place? **[1]**

(b) Outline the processes used to produce ethene from crude oil. **[4]**

(c) Write down one use for ethene apart from making ethanol. **[1]**

(d) Write word and symbol equations for the reaction producing ethanol from ethene. **[3]**

(e) Low-grade wine containing ethanol can be turned into vinegar in the equipment shown in the diagram shown below.

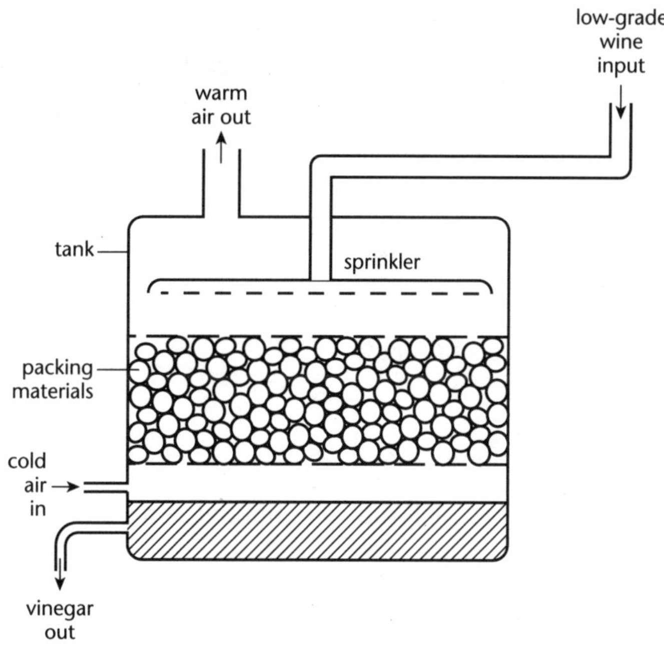

The wine is sprayed downward and air passes upwards over a packaging material. Bacteria, growing on the packing material, convert the ethanol in the wine into ethanoic acid.

The equation for the oxidation of ethanol into ethanoic acid is:

$C_2H_5OH(aq) + O_2(g) \rightarrow CH_3COOH(aq) + H_2O(l)$

(i) Why is this reaction an oxidation reaction? **[1]**

(ii) What evidence is there that the reaction is exothermic? **[1]**

Exam practice questions

2. Aspartame is an artificial sweetener.

Its structure is shown below.

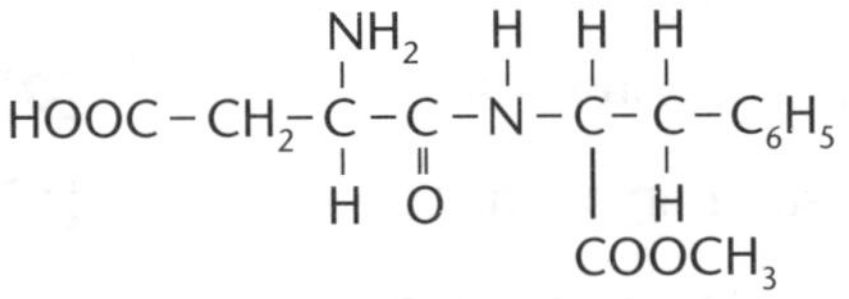

$$HOOC-CH_2-\underset{H}{\overset{NH_2}{C}}-\underset{O}{\overset{}{C}}-\overset{H}{N}-\underset{COOCH_3}{\overset{H}{C}}-\underset{H}{\overset{H}{C}}-C_6H_5$$

(a) What is the molecular formula of aspartame? **[1]**

(b) Is aspartame a carbohydrate? Explain your answer **[1]**

(c) Why is aspartame better than sugar for a person on a reduced calorie diet? **[1]**

3. Glucose can be represented as

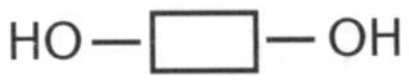

HO—☐—OH

Draw the structure of a more complex carbohydrate that can be formed from glucose by condensation polymerisation. **[2]**

4. Esters occur naturally in plants and animals. They are also manufactured from petroleum. Ethyl ethanoate and butyl ethanoate are industrially important as solvents.

(a) **(i)** Explain the term 'solvent'. **[1]**

(ii) Give the formula of ethyl ethanoate. **[1]**

(iii) Ethyl ethanoate can be made from ethanol and ethanoic acid. Describe how the following chemicals can be made.

- ethanol from ethene **[2]**
- ethanoic acid from ethanol **[2]**

(iv) Name two chemicals from which methyl ethanoate can be made. **[1]**

(b) The following equation represents the alkaline hydrolysis of a naturally occurring ester.

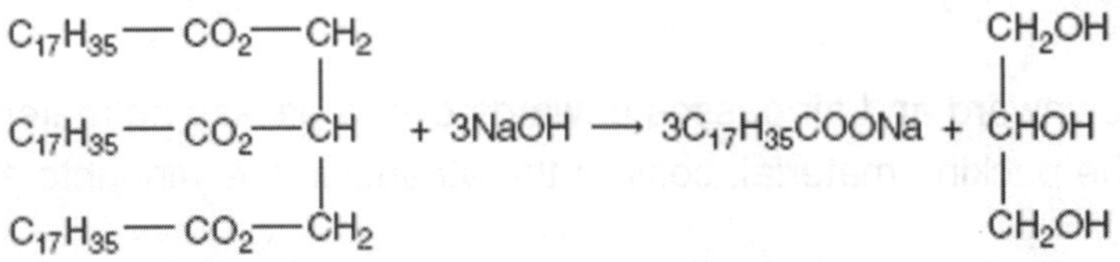

$$\begin{array}{l} C_{17}H_{35}-CO_2-CH_2 \\ \qquad\qquad\qquad\quad | \\ C_{17}H_{35}-CO_2-CH \\ \qquad\qquad\qquad\quad | \\ C_{17}H_{35}-CO_2-CH_2 \end{array} + 3NaOH \longrightarrow 3C_{17}H_{35}COONa + \begin{array}{l} CH_2OH \\ | \\ CHOH \\ | \\ CH_2OH \end{array}$$

(i) Which substance in the equation is an alcohol? Underline the substance in the equation above. **[1]**

(ii) What is the major use for compounds of the type $C_{17}H_{35}COONa$? **[1]**

Chapter 4

Carbonates

The following topics are covered in this section:

- ***Products made from rocks***

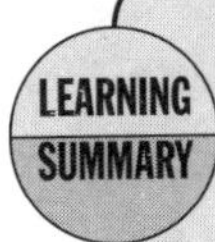
LEARNING SUMMARY

After studying this section you should be able to:

- ***recall some materials made from rocks and the uses of these materials***
- ***understand how salt and limestone can be used to make useful materials.***

Products made from rocks

The rocks of the Earth are the source of a wide range of materials.

Rocks as building materials

Rocks such as **limestone**, **sandstone** and **slate** are used as **building materials**. As transport costs are very high, rocks are often used as building materials close to where they are dug out of the ground **(quarried)**.

Building materials made from rocks

As natural rocks are expensive, new materials have been developed to replace them. **Table 4.1** summarises how some of these materials are made.

Table 4.1 Building materials made from rocks

Material	How it is made	More information
Bricks	*By baking clay to a high temperature*	*Hard and brittle – regular shape.*
Mortar	*Mixture of* **calcium hydroxide**, **sand** *and* **water** *made into a thick paste*	*It sets by losing water and by absorbing carbon dioxide from the air. Long crystals of calcium carbonate are formed.*
Cement	*Heating* **limestone** *with* **clay** *(containing aluminium and silicates)*	*It consists of a complex mixture of calcium and aluminium silicates. When it is mixed with water, a chemical reaction occurs producing calcium hydroxide, and this sets in a similar way to mortar.*
Concrete	*Made by mixing cement with sand and small stones*	*Used to make many objects such as drain covers that were previously made from cast iron. Concrete can be strengthened by steel reinforcing rods.*

(continued...)

Table 4.1 *(continued...)*

Material	How it is made	More information
Glass	*Mixing* **limestone**, **sand** *(silicon dioxide) and* **sodium carbonate** *together and melting the mixture*	*The resulting mixture of calcium and sodium silicates cools to produce glass. Coloured glass is due to transition metal oxides present in the mixture*

Chemicals from rocks

Limestone is the mineral extracted from the Earth in the largest amounts. Often, it is found in beautiful areas and its mining can damage the environment.

Many chemicals are made from rocks.

These processes are summarised in **Fig. 4.1**.

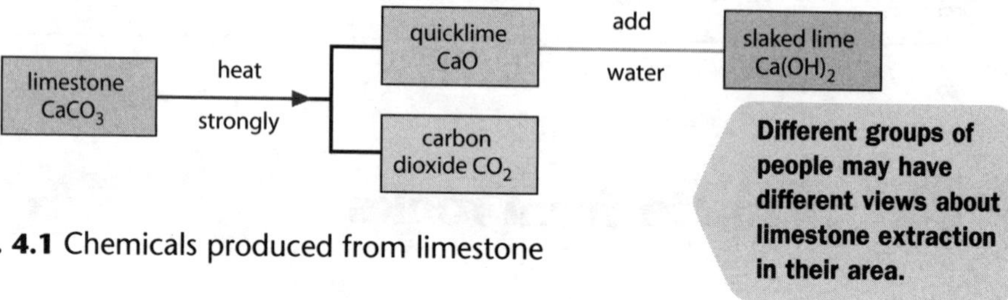

Fig. 4.1 Chemicals produced from limestone

Different groups of people may have different views about limestone extraction in their area.

Uses of Limestone

Apart from its use in building construction, limestone has several other uses, some of which are listed below.

- Neutralising acidic soil by mixing soil with powdered limestone.
- In the manufacture of iron, limestone is used to remove impurities like sand by forming slag.
- It is used to manufacture sodium carbonate, the key ingredient for making soaps, detergents etc.
- Used in manufacture of cement (Ref. Table 4.1).

KEY POINT **Limestone is used to make quicklime (calcium oxide) and slaked lime (calcium hydroxide).**

Quicklime and **slaked lime** have several uses, some of which are listed below:

- Neutralising acidic soil by mixing it with powdered lime.
- Improving drainage properties of clayey soils – when mixed with powdered lime, their clayey nature decreases.
- Used as a water-softening agent and also to manufacture bleaching powder.
- Used for making mortar in building construction.
- Used for whitewashing and colourwashing.
- Used in the manufacture of paper and glass, in leather tanning and sugar refining.
- Limewater(solution of slaked lime in water) is used in medicine as an antacid, as a neutraliser for acid poisoning and for treatment of burns.
- Used for neutralising acidic industrial waste products.

KEY POINT

Rock salt is used as the raw material for producing a wide range of chemicals including sodium hydroxide, chlorine, hydrogen, sodium and household bleach.

These processes are summarised in **Fig. 4.2**.

Salt is mined by solution mining. Water is pumped underground. The salt dissolves and salt solution (brine) is pumped to the surface.

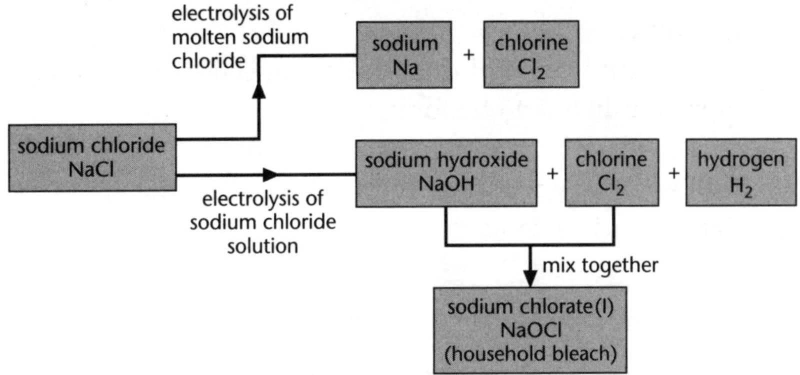

Fig. 4.2 Chemicals produced from rock salt

Metals from rocks

Most metals are found in the earth as deposits of **ore**.

KEY POINT

An ore is a rock that contains enough of a metal compound for it to be worth extracting the metal

Here, metals are arranged in the order of their reactivity. Notice that most reactive metals are present as chlorides,carbonates or oxides and less reactive metals as sulphides.

Table 4.2 Common metal ores and their chief compound

Metal	Name of ore	Compound of metal present
Sodium	*Rock salt (halite)*	*Sodium chloride*
Magnesium	*Magnesite*	*Magnesium chloride*
Aluminium	*Bauxite*	*Aluminium oxide*
Iron	*Haematite*	*Iron(III) oxide*
Zinc	*Zinc blende*	*Zinc sulphide*
Mercury	*Cinnabar*	*Mercury(II) sulphide*

You will certainly need to know the names of the ores of aluminium and iron.

Some ores contain only small amounts of metal compounds. The metal compound in these ores may be concentrated by **froth flotation** before the metal is extracted. The ore is added to a detergent bath and the mixture agitated. By careful control of the conditions, it is possible to get the metal compound to float while the impurities sink to the bottom.

Sample IGCSE questions

1. What are the uses of limestone? **[2]**

Neutralising acidic soil by mixing soil with powdered limestone.
In the manufacture of iron, limestone is used to remove impurities like sand by forming slag
Used to manufacture sodium carbonate
Used as the key ingredient for making soaps, detergents etc.
Manufacture of cement.

You may include any two of these given uses in your answer.

2. What are the uses of quick lime and slaked lime? **[2]**

To neutralise acidic soil - by mixing soil with powdered lime,
To improve drainage properties of clayey soils - when mixed with powdered lime their clayey nature decreases,
Used as a water-softening agent
To manufacture bleaching powder,
Used for making mortar in building construction.

You may include any two of these given uses in your answer.

Exam practice questions

1. Which two chemicals can be used to reduce acidity of soil? **[2]**
2. Which gas is evolved on heating limestone? **[1]**
3. What are the various chemicals that can be produced from limestone? **[3]**

Chapter 5

Stoichiometry

The following topics are covered in this section:

- **Stoichiometry**
- **The mole concept**

5.1 Stoichiometry

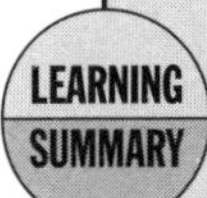

LEARNING SUMMARY

After studying this section you should be able to:

- ***use the symbols of elements and write formulae of simple compounds***
- ***write chemical formulae and balanced symbol equations***
- ***calculate quantities of chemicals reacting or produced***
- ***use masses to calculate the simplest formula for a compound.***

The study of the numbers of atoms combining to form a chemical compound is called **stoichiometry**. A knowledge of the amounts of these elements that combine, together with a knowledge of their atomic weights, yield their **atomic numbers**.

Symbols of Elements

For ease of use, the elements have been given shorthand symbols. These symbols have been created mostly from the element names themselves, by combining the first letter of the element's name with some other letter of the same, e.g. Ca is used for Calcium. Some symbols have also been derived from the latin names of these elements, e.g. Gold is denoted by the symbol Au (Latin word for gold is *Aurum*). The symbols consist of one, two or three letters, the first of which is capital. A list of some common elements and their symbols is given in **Table 5.1**.

Table 5.1 Elements and their symbols

Element	Latin Name	Symbol	Physical state at room temperature and pressure
Aluminium		*Al*	*Solid*
Argon		*Ar*	*Gas*
Barium		*Ba*	*Solid*
Boron		*B*	*Solid*
Bromine		*Br*	*Liquid*
Calcium		*Ca*	*Solid*
Carbon		*C*	*Solid*
Chlorine		*Cl*	*Gas*
Chromium		*Cr*	*Solid*
Copper	*(Cuprum)*	*Cu*	*Solid*
Fluorine		*F*	*Gas*

(continued...)

Table 5.1 Elements and their symbols *(continued...)*

Element	Latin Name	Symbol	Physical state at room temperature and pressure
Germanium		Ge	Solid
Gold	(Aurum)	Au	Solid
Helium		He	Gas
Hydrogen		H	Gas
Iodine		I	Solid
Iron	(Ferrum)	Fe	Solid
Lead	(Plumbum)	Pb	Solid
Magnesium		Mg	Solid
Mercury	(Hydragyrum)	Hg	Liquid
Neon		Ne	Gas
Nitrogen		Ne	Gas
Oxygen		O	Gas
Phosphorus		P	Solid
Potassium	(Kalium)	K	Solid
Silicon		Si	Solid
Silver	(Argentum)	Ag	Solid
Sodium	(Natrium)	Na	Solid
Sulphur		S	Solid
Tin	(Stannum)	Sn	Solid
Zinc		Zn	Solid
Note : For complete list of elements see the periodic table.			

Formulae of Simple Compounds

Compounds are pure substances which are formed when two or more elements combine together chemically. The symbols of elements can be used to depict the substances so formed, e.g. since one oxygen atom combines with two atoms of hydrogen to form water so the chemical formula of water is H_2O.

Names and formulae of some common compounds are enlisted in **Table 5.2**.

Table 5.2 Formulae of some common compounds

Compound	Formula
Ammonia	NH_3
Calcium Hydroxide	$Ca(OH)_2$
Carbon Monoxide	CO
Carbon Dioxide	CO_2
Copper Sulphate	$CuSO_4$

(continued...)

Table 5.2 Formulae of some common compounds *(continued...)*

S.No.	Compound	Formula
6	*Hydrochloric Acid*	HCl
7	*Nitric Acid*	HNO_3
8	*Sodium Carbonate*	Na_2CO_3
9	*Sodium Hydroxide*	$NaOH$
10	*Sulphuric Acid*	H_2SO_4

The chemical formulae of compounds can be best understood from the following diagrammatic representation:

Fig 5.1 The pure element hydrogen reacts with the pure element oxygen to form the compound water.

In **Fig. 5.1** it is shown how hydrogen (a pure element), when combined with oxygen (another pure element) forms water – a pure compound. This new compound is formed as a result of hydrogen, burning in oxygen, and the water molecule thus formed contains two atoms of hydrogen and one atom of oxygen.

Equations

Word equations

A chemical reaction can be summarised by an **equation**.The simplest equation is a **word equation**:

e.g. when **sodium hydroxide** and **hydrochloric acid** react, **sodium chloride** and **water** are formed.

sodium hydroxide + hydrochloric acid → sodium chloride + water

The substances on the left-hand side (sodium hydroxide and hydrochloric acid, in this case) are called **reactants**. The substances produced (sodium chloride and water) are called **products**.

Although word equations may be useful, they do not give a full picture of what is happening.

Symbol equations

Reactions can be summarised using **chemical symbols**. This is a system which is used throughout the world.

The equation for the reaction between sodium hydroxide and hydrochloric acid is written as:

$NaOH + HCl \rightarrow NaCl + H_2O$

This equation is correctly **balanced**, i.e. there are the same number of each type of atom on each side of the equation.

Calcium hydroxide reacts with hydrochloric acid to form calcium chloride and water. The **formula** for calcium hydroxide is not CaOH but $Ca(OH)_2$ (see page 63) and the formula of calcium chloride is not CaCl but $CaCl_2$.

The equation can be written:

$Ca(OH)_2 + HCl \rightarrow CaCl_2 + H_2O$

This equation is **unbalanced** because there are different numbers of atoms on each side:

Notice that the small number after the bracket multiplies everything in the bracket.

Left-hand side		Right-hand side	
1	Ca	1	Ca
2	O	1	O
3	H	2	H
1	Cl	2	Cl

It is an important law, called the **law of conservation of mass**, that atoms cannot be made or destroyed during a chemical reaction. They can just be rearranged.

You cannot write:

$CaOH + HCl \rightarrow CaCl + H_2O$

This would mean altering the formulae which is not permissible.

Instead, you have to change the proportions of these substances by altering the large numbers at the front:

$Ca(OH)_2 + 2HCl \rightarrow CaCl_2 + 2H_2O$

Left-hand side		Right-hand side	
1	Ca	1	Ca
2	O	2	O
4	H	4	H
2	Cl	2	Cl

Always check whether an equation is balanced before moving on. Leaving An equation unbalanced loses you a mark.

State symbols

Sometimes state symbols are added to symbol equations to show whether the substance is solid, liquid or gas or whether it is in solution.

There is usually no penalty if you miss out state symbols unless the question specifically asks you to add them.

These state symbols are:

(s) solid

(l) liquid

(g) gas

(aq) in aqueous solution, i.e. where the solvent is water.

An example of an equation with state symbols is:

$2Na(s) + 2H_2O(l) \rightarrow 2NaOH(aq) + H_2(g)$.

Ionic equations

Consider the reaction of sodium hydroxide and hydrochloric acid again. Sodium hydroxide, hydrochloric acid and sodium chloride are made up of ions.

The equation could be written as:

$Na^+ OH^- + H^+ Cl^- \rightarrow Na^+ Cl^- + H_2O$.

Since an equation shows change, anything which appears unchanged on both sides of the equation can be removed.

The simplest equation is

$OH^- + H^+ \rightarrow H_2O$.

This ionic equation, in addition to having the same number of each type of atom on each side, also has **equal charge on each side**. In this case, the sum of the charges on each side is zero.

When you write an ionic equation, check that no ions appear on both sides. These ions called 'spectator ions' and can be missed out. The net equation shows only a change.

When chlorine is bubbled through potassium iodide solution, potassium chloride and iodine are produced.

The symbol equation is:

$2KI + Cl_2 \rightarrow 2KCl + I_2$.

Potassium iodide and potassium chloride are made up of ions.

$2K^+ I^- + Cl_2 \rightarrow 2K^+ Cl^- + I_2$

The ionic equation is

$2\,I^- + Cl_2 \rightarrow 2\,Cl^- + I_2$.

Now there are two iodines and two chlorines on each side and the charge on each side is –2.

PROGRESS CHECK

Balance each of the following equations.

1. $Mg + HCl \rightarrow MgCl_2 + H_2$
2. $Na + Cl_2 \rightarrow NaCl$
3. $H_2 + O_2 \rightarrow H_2O$
4. $H_2O_2 \rightarrow H_2O + O_2$
5. $H_2 + Cl_2 \rightarrow HCl$
6. $NO + O_2 \rightarrow NO_2$
7. $O^{2-} + H^+ \rightarrow H_2O$
8. $Na + H^+ \rightarrow Na^+ + H_2$

1. $Mg + \mathbf{2}HCl \rightarrow MgCl_2 + H_2$; 2. $\mathbf{2}Na + Cl_2 \rightarrow \mathbf{2}NaCl$;
3. $\mathbf{2}H_2 + O_2 \rightarrow \mathbf{2}H_2O$; 4. $\mathbf{2}H_2O_2 \rightarrow \mathbf{2}H_2O + O_2$
5. $H_2 + Cl_2 \rightarrow \mathbf{2}HCl$; 6. $\mathbf{2}NO + O_2 \rightarrow \mathbf{2}NO_2$
7. $O^{2-} + \mathbf{2}H^+ \rightarrow H_2O$; 8. $\mathbf{2}Na + \mathbf{2}H^+ \rightarrow \mathbf{2}Na^+ + H_2$

Relative atomic mass and relative formula mass

Atoms are too small to be weighed individually. It is possible, however, to compare the mass of one atom with the mass of another.

This is done using a **mass spectrometer**. For example, a magnesium atom has twice the mass of a carbon-12 atom and six times the mass of a helium atom.

KEY POINT **The relative atomic mass of an atom is the number of times an atom is heavier than one-twelfth of a carbon-12 atom.**

Relative atomic masses are not all whole numbers because of the existence of isotopes (see 2.1).

You are not expected to remember relative atomic masses. They are given on examination papers in one of the following ways:
1. $A_r(Ca) = 40$ or
2. Ca = 40

The **relative atomic mass (A_r)** is simply a **number** and has **no units**.

For compounds, if you know the formula, you can use relative atomic masses to work out **relative formula masses (M_r)**, e.g. Work out the relative formula mass of water, H_2O, given $A_r(H) = 1$ and $A_r(O) = 16$

The relative formula mass of water
$M_r = (2 \times 1) + 16 = 18.$

The relative formula mass can be worked out by adding relative atomic masses.

Using equations to calculate masses

A balanced symbol equation can tell you about the chemicals involved in a reaction as reactants or products. It can also tell you about the **masses** of chemicals which react or are formed in the reaction.

In order to do this, you need **relative atomic masses** of the elements.

The relative atomic masses can be looked up in a data book or obtained from the periodic table.

Iron and sulphur react together to form iron(II) sulphide.

The symbol equation is:

$Fe + S \rightarrow FeS$

Notice that the sum of the masses of the reactants equals the mass of the product.

The relative atomic mass of iron is 56 and that of sulphur is 32.

From the equation, using relative atomic masses

56 g of iron combine with 32 g of sulphur to form 88 g of iron(II) sulphide.

Another example:

Carbon burns in excess oxygen to form carbon dioxide.

Calculate the mass of carbon dioxide produced when 1 g of carbon is burned.

($A_r(C) = 12$, $A_r(O) = 16$)

Always check that the mass of the reactants is the same as the mass of the products.

First write the symbol equation:

$C + O_2 \rightarrow CO_2$.

Now use relative atomic masses to work out masses of reactants and products.

12 g of carbon react with (2 × 16) g of oxygen to form (12 + (2 × 16)) g of carbon dioxide.

Thus, 12 g of carbon react with 32 g of oxygen to form 44 g of carbon dioxide.

If 1 g of carbon is used (one-twelfth of the quantity), the mass of carbon dioxide formed would be one-twelfth, i.e. 44/12 = 3.7 g.

($A_r(H) = 1$, $A_r(C) = 12$, $A_r(O) = 16$, $A_r(Mg) = 24$, $A_r(S) = 32$, $A_r(K) = 39$

1. *How many times heavier is a magnesium atom than a carbon atom?*
2. *What is the relative formula mass of methane, CH_4?*
3. *What is relative formula mass of sulphuric acid, H_2SO_4?*
   The equation for the action of heat on potassium hydrogencarbonate
   $2KHCO_3 \rightarrow K_2CO_3 + H_2O + CO_2$
4. *What is the relative formula mass of potassium hydrogencarbonate?*
5. *What is the relative formula mass of potassium carbonate?*
6. *What mass of potassium carbonate would be formed when 10 g of potassium hydrogencarbonate are completely decomposed?*

1. Twice; 2. 16; 3. 98; 4. 100; 5. 138; 6. 13.8 g

Working out chemical formulae

Chemical formulae can be worked out using the formulae of common ions.

Table 5.3 contains some of the common positive and negative ions.

Positive ions			**Negative ions**		
+1	+2	+3	−1	−2	−3
sodium Na^+	*magnesium* Mg^{2+}	*aluminium* Al^{3+}	*chloride* Cl^-	*sulphate* SO_4^{2-}	*phosphate* PO_4^{3-}
potassium K^+	*calcium* Ca^{2+}		*nitrate* NO_3^-	*carbonate* CO_3^{2-}	
hydrogen H^+	*lead* Pb^{2+}		*hydroxide* OH^-	*oxide* O^{2-}	
ammonium NH_4^+	*zinc* Zn^{2+}				
silver Ag^+	*copper* Cu^{2+}				

If you want to write a chemical formula, you will need to use the correct ions:

Remember metals form positive ions.

e.g. **Sodium chloride** Na^+ and Cl^-

As there are **equal numbers of positive and negative charges**, you can write the formula as NaCl.

Sodium sulphate Na^+ and SO_4^{2-}

There are **twice as many negative charges as positive charges**.

In the formula, there need to be twice as many sodium ions.

The formula is therefore written as Na_2SO_4.

Aluminium oxide Al^{3+} and O^{2-}

In order to get equal numbers of positive and negative charges, you have to take two aluminium ions for every three oxide ions. The formula is Al_2O_3.

It is possible to work out the formula of a compound, using results from an experiment.

Magnesium oxide

If magnesium burns in oxygen, magnesium oxide is formed.

Here are the results of an experiment.

Mass of crucible + lid = 25.15 g

Mass of crucible + lid + magnesium = 25.27 g

∴ Mass of magnesium = 25.27 – 25.15 = 0.12 g

Mass of crucible + lid + magnesium oxide = 25.35 g

∴ Mass of magnesium oxide = 25.35 – 25.15 g = 0.20 g

From these results

0.12 g of magnesium combines with (0.20 – 0.12 g) of oxygen to form 0.20 g of magnesium oxide.

0.12 g of magnesium combines with 0.08 g of oxygen.

Divide each mass by the appropriate relative atomic mass:
$A_r(Mg) = 24$, $A_r(O) = 16$.

Magnesium $\frac{0.12}{24} = 0.05$ Oxygen $\frac{0.08}{16} = 0.05$

Divide by the smallest number (here they are both the same)

1 1

The simplest formula is MgO.

Lead oxide

4.14 g of lead combines with 0.64 g of oxygen.

$A_r(Pb) = 207$, $A_r(O) = 16$

Divide by the appropriate relative atomic masses

Lead $\frac{4.14}{207}$ 0.02 Oxygen $\frac{0.64}{16}$ 0.04

Divide by the smallest, i.e. 0.02

1 2

The simplest formula for this lead oxide is PbO_2.

A common mistake here is to write the formula as Pb_2O.

PROGRESS CHECK

$A_r(H) = 1$, $A_r(C) = 12$, $A_r(N) = 14$, $A_r(O) = 16$, $A_r(Cu) = 64$

1.6 g of copper oxide produces 1.28 g of copper.

1. *What mass of oxygen combines with 1.28 g of copper?*
2. *Choose the formula of this copper oxide from the list:*
   Cu_2O CuO CuO_2
3. *6 g of carbon combines with 1 g of hydrogen.*
   *Choose the **simplest** formula of this compound.*
   CH_2 C_2H_4 CH_4

0.7 g of nitrogen combines to form 1.5 g of nitrogen oxide.

4. *What mass of oxygen combines with 0.7 g of nitrogen?*
5. *Choose the simplest formula of this compound.*
   N_2O NO_2 NO

1. 0.32 g; 2. CuO; 3. CH_2; 4. 0.8 g; 5. NO

5.2 The mole concept

LEARNING SUMMARY

After studying this section you should be able to:

- ***recall that mole is the amount of substance containing Avogadro's number of particles***
- ***explain that one mole of any substance contains the same number of particles***
- ***work out the number of moles present in a given mass of substance***
- ***work out the mass in g moles of different substances***
- ***work out formulae from percentages and percentages from formulae***
- ***work out volume changes during reactions involving gases***
- ***work out concentrations of solutions in mol/dm³.***

The mole

KEY POINT

The relative atomic mass of an atom is the number of times an atom is heavier than one twelfth of a carbon-12 atom. (See 5.1)

As an alternative to consider masses of individual atoms, it is possible to compare a large numbers of atoms.

Atoms are so small that individual atoms cannot be weighed separately.

The relative atomic mass of magnesium is twice the relative atomic mass of carbon.

1 atom of magnesium weighs twice as much as 1 atom of carbon-12. 2 atoms of magnesium weigh twice as much as 2 atoms of carbon-12. 100 atoms of magnesium weigh twice as much as 100 atoms of carbon-12.

KEY POINT

The mass of magnesium atoms will always be twice the mass of the carbon-12 atoms, provided equal numbers of atoms are compared.

We frequently use collective terms to describe a number of objects, e.g. a dozen eggs, a gross of test tubes, etc. In chemistry, the term **mole** (abbreviation mol) is used in the same way.

We refer to a mole of magnesium atoms, a mole of carbon dioxide molecules or a mole of electrons.

A mole is that amount of matter which contains 6×10^{23} particles (600 000 000 000 000 000 000 000). This number is called Avogadro's constant (L).

Avogadro's number is so large that if the whole population of the world wished to count up to this number between them and they worked at counting without breaks it would take six million years to finish.

A mole of atoms of any element has a mass equal to the relative atomic mass (but with units of grams).

One mole of magnesium atoms has a mass of 24g and one mole of carbon atoms has a mass of 12g. Notice that the mass of magnesium is still twice the mass of carbon, because equal numbers of particles are considered.

The mole may be defined as the amount of substance which contains as many elementary units as there are atoms in 12g of carbon.

A mole of carbon atoms is a small handful. As there are 6×10^{23} atoms in the pile, it emphasises just how small atoms are.

These elementary units can be considered as:

atoms e.g. Mg, C, He

molecules e.g. CH_4, H_2O

ions e.g. Na^+, Cl^-

specified formula units e.g. H_2SO_4

The term '1 mole of chlorine' can be ambiguous. It could mean 1 mole of chlorine atoms (6×10^{23} atoms) or 1 mole of chlorine molecules, Cl_2 (12×10^{23} atoms).

Calculating the mass of 1 mole

Calculate the mass of:

(a) 1 mole of oxygen atoms, O

(b) 1 mole of oxygen molecules, O_2

(c) 1 mole of methane molecules, CH_4

(d) 1 mole of sulphuric acid, H_2SO_4.

If you look up relative atomic masses on the periodic table make sure you use relative atomic masses rather than atomic numbers.

Relative atomic masses $A_r(H) = 1$, $A_r(C) = 12$, $A_r(O) = 16$, $A_r(S) = 32$.

(a) 16 g

This is the relative atomic mass of oxygen in g.

(b) $16 \times 2 = 32$ g

(c) $12 + (4 \times 1) = 16$ g

(d) $(2 \times 1) + 32 + (4 \times 16) = 98$ g

Calculating the number of moles in given masses

$$\textbf{number of moles} = \frac{\textbf{mass in g}}{\textbf{mass of 1 mole in g}}$$

Calculate the number of moles represented by:

(a) 8 g of oxygen molecules

(b) 8 g of methane molecules

(c) 9.8 g of sulphuric acid

(d) 22 g of carbon dioxide, CO_2.

(a) Number of moles = $\frac{8}{32}$ = 0.25

(b) Number of moles = $\frac{8}{16}$ = 0.5

(c) Number of moles = $\frac{9.8}{98}$ = 0.1

(d) Mass of 1 mole of CO_2 = 12 + (2 × 16) = 44 g

Number of moles = $\frac{22}{44}$ = 0.5

This formula can be rearranged

mass in g = number of moles × mass of 1 mole in g.

Calculate the mass of:

(a) 0.1 moles of methane

(b) 2 moles of sulphuric acid

(c) 0.5 moles of sulphur dioxide, SO_2.

(a) Mass = 0.1 × 16 = 1.6 g

(b) Mass = 2 × 98 = 196 g

(c) Mass of 1 mole of sulphur dioxide = 32 + (2 × 16) = 64 g

Mass = 0.5 × 64 = 32 g.

Volume of 1 mole of gas

There is no simple relationship that predicts the volume occupied by 1 mole of molecules in a solid or liquid.

For gases

This information will be given on examination papers if it is needed.

1 mole of any gas occupies 24 000 cm^3 (or 24 dm^3) at room temperature and atmospheric pressure.

Calculate the volume of 0.1 moles of carbon dioxide at room temperature and atmospheric pressure.

Answer = 24 000 × 0.1 = 2400 cm^3.

PROGRESS CHECK

Relative atomic masses: Ar(H) = 1, Ar(C) = 12, Ar(N) = 14, Ar(O) = 16, Ar(Al) = 27; Ar(Ca) = 40; Ar(Fe) = 56; Ar(Pb) = 207

Write down the number of moles of atoms in:

1. *60 g of carbon;*
2. *3 g of aluminium;*
3. *40 g of iron(III) oxide, Fe_2O_3;*
4. *1 g of calcium carbonate, $CaCO_3$;*
5. *0.2 g of hydrogen molecules, H_2.*

What are the masses of the following?

6. *10 moles of water, H_2O;*
7. *0.5 moles of ammonium nitrate, NH_4NO_3;*
8. *2 moles of ethanol, C_2H_5OH;*
9. *0.01 moles of lead(II) nitrate, $Pb(NO_3)_2$.*
10. *What is the volume of 11 g of carbon dioxide at room temperature and atmospheric pressure?*

1. 5; 2. 0.11; 3. 0.25; 4. 0.01; 5. 0.1; 6. 180 g; 7. 40 g; 8. 92 g; 9. 3.31 g;
10. 6000 cm^3

Calculating formulae from percentages

Chemical formulae were worked out using relative atomic masses.

Chemical formulae can be worked out using moles.

e.g. 4.14 g of lead combines with 0.64g of oxygen. (A_r(Pb) = 207; A_r(O) =16)

Number of moles of lead = 4.14 ÷ 207 = 0.02

Number of moles of oxygen = 0.64 ÷ 16 = 0.04

A common error here is to write Pb_2O.

There are twice as many particles of oxygen as lead, so the simplest formula is PbO_2.

You can use percentages of different elements to work out the formula.

Calculate the formula of a compound containing iron, sulphur and oxygen. It contains 28% iron and 24% sulphur.

Remember that the percentages of all the elements add up to 100

Relative atomic masses : A_r(Fe) = 56; A_r(S) = 32; A_r(O) = 16.

Percentage of oxygen in the compound = 100 – (28 + 24) = 48%.

Number of moles of iron = 28 ÷ 56 = 0.5

Number of moles of sulphur = 24 ÷ 32 = 0.75

Number of moles of oxygen = 48 ÷ 16 = 3

0.25 divides into 0.5, 0.75 and 3.

Divide each by 0.25.

The simplest formula is $Fe_2S_3O_{12}$.

The simplest formula is called the **empirical formula**.

KEY POINT **The molecular formula is either the empirical formula or some multiple of it.**

The empirical formula of a compound is CH_2. The mass of 1 mole of the compound is 56 g.

The molecular formula of the compound is 4 × the empirical formula i.e. C_4H_8.

This has a mass of: 4 × 14 = 56 g.

Calculate the percentage of an element in a compound (%yield and % purity)

Calculate the percentage of potassium in potassium hydrogencarbonate, $KHCO_3$.

Relative atomic masses A_r(H) = 1, A_r(C) = 12, A_r(O) = 16 , A_r(K) = 39.

These calculations are often used to calculate percentage of nitrogen, phosphorus or potassium in a fertiliser.

Mass of 1 mole of potassium hydrogencarbonate = 39 + 1 + 12 + (3×16) = 100 g.

Percentage of potassium = $\frac{39}{100} \times 100$ = 39%.

PROGRESS CHECK

A hydrocarbon X contains 75% carbon.

1. *Calculate the percentage of hydrogen in the hydrocarbon.*
2. *Calculate the empirical (simplest) formula of the hydrocarbon.*

Relative atomic masses $A_r(H) = 1$, $A_r(C) = 12$.

3. *The mass of 1 mole of the hydrocarbon is 16g. What is the molecular formula?*
4. *Suggest a hydrocarbon that could be X.*
5. *What is the percentage of nitrogen in ammonium nitrate,* NH_4NO_3*?*
   Relative atomic masses $A_r(H) = 1$, $A_r(N) = 14$, $A_r(O) = 16$.

1. 25%; 2. CH_4; 3. CH_4; 4. Methane; 5. 35%.

Volume changes in chemical reactions

On page 62 (5.1) it was shown how masses of chemicals reacting, and masses of products formed could be worked out using a balanced symbol equation.

e.g. The equation for the burning of carbon in excess oxygen is

$$C(s) + O_2(g) \rightarrow CO_2(g).$$

The equation shows us that 1 mole of carbon (12 g) reacts with 1 mole of oxygen (32 g) to produce 1 mole of carbon dioxide.

Remember the sum of the mass of the reactants is the same as the sum of the mass of the products. The number of moles on each side is not the same.

We know that

KEY POINT **1 mole of molecules of any gas occupies 24 000 cm³ (or 24 dm³) at room temperature and atmospheric pressure.**

12g of carbon react with 24 dm^3 of oxygen (at room temperature and atmospheric pressure) to form 24 dm^3 of carbon dioxide.

Another example:

hydrogen + oxygen → water

$$2H_2(g) + O_2(g) \rightarrow 2H_2O(l).$$

4g of hydrogen reacts with 32g of oxygen to form 36g of water.

In terms of moles

2 moles of hydrogen molecules reacts with 1 mole of oxygen molecules to form 2 moles of water.

At room temperature and atmospheric pressure

48 dm^3 of hydrogen reacts with 24 dm^3 of oxygen. As steam has condensed to liquid water, the volume of the products is negligible.

There is a huge decrease in volume during the reaction.

KEY POINT **The sum of the volumes of the reactants is not necessarily equal to the sum of the volumes of the products.**

PROGRESS CHECK

1. *In which of these reactions is the volume of the reactants equal to the volume of the products?*
   A. $CH_4(g) + 2O_2(g) \rightarrow CO_2(g) + 2H_2O(l)$
   B. $H_2(g) + Cl_2(g) \rightarrow 2HCl(g)$
   C. $2CH_4(g) + 3O_2(g) \rightarrow 2CO(g) + 4H_2O(l)$
   D. $N_2(g) + 3H_2(g) \rightarrow 2NH_3(g)$

The equation for the complete combustion of ethane is

$$2C_2H_6(g) + 7O_2(g) \rightarrow 4CO_2(g) + 6H_2O(l)$$

2. *What volume of oxygen is needed to react with 30 cm³ of ethane, C_2H_6?*
3. *What volume of carbon dioxide would be produced by the complete combustion of 20 cm³ of ethane.*

1. B; 2. 105 cm³; 3. 40 cm³.

Concentration of solutions

Measuring the concentration of a solution

The concentration of a solution can be measured in units of g/dm³.

For example, if 10 g of salt are dissolved in 100 g of water, the concentration of the salt solution is 100 g/dm³.

g/dm³ and g/litre are the same. Sometimes they are written as g dm^{-3} or g l^{-3}.

Measuring concentration in this way gives no comparison of the number of particles in a given volume of solution.

When 1 mole of solute is dissolved in water and the solution made up to a volume of 1 dm³ the solution has a concentration of 1 mole/dm³. This is usually written as 1 mol/dm³ and sometimes called a molar (or M) solution.

Equal volumes of two solutions each 1 mol/dm³ will contain the same number of particles.

e.g. 8 g of sodium hydroxide solution is dissolved in water and made up to a volume of 100 cm³ with water.

(a) What is the concentration of the solution in mol/dm³?

(b) How many moles of sodium hydroxide are present in 25 cm³ of this solution?

Relative atomic masses: $A_r(H) = 1$ $A_r(O) = 16$, $A_r(Na) = 23$.

(a) 1 mole of NaOH = 23 + 16 + 1 = 40 g

Number of moles of NaOH = 8 ÷ 40 = 0.2 moles

Concentration of the solution is 0.2 moles/100 cm³ or 2 mol/dm³

(b) 25 cm³ of solution is one fortieth of one dm³, so it contains 2 ÷ 40 moles i.e. 0.05 moles.

PROGRESS CHECK

1. *Four solutions of potassium hydroxide are:*
   (a) 100 cm^3 of 1 mol/dm^3 *(c) 50 cm^3 of 2 mol/dm^3*
   (b) 10 cm^3 of 10 mol/dm^3 *(d) 10 000 cm^3 of 0.01 mol/dm^3*
   What do these solutions of potassium hydroxide have in common?
2. *Calculate the concentration (in mol/dm^3) of a solution of sulphuric acid, H_2SO_4, containing 9.8 g in 100 g of solution*
   Relative atomic masses: $A_r(H) = 1$, $A_r(O) = 16$, $A_r(S) = 32$.

Number of moles present = concentration (in mol/dm^3) $\times \frac{\text{volume in cm}^3}{1000}$

Work out the number of moles of hydrochloric acid present in:

3. *500 cm^3 of solution 2 mol/dm^3*
4. *5000 cm^3 of solution 0.2 mol/dm^3*
5. *100 cm^3 of solution 5 mol/dm^3.*

1. They all contain the same number of moles of potassium hydroxide (0.1 moles) and so the same number of particles; 2. 1 mol/dm^3; 3. 1 mol/dm^3; 4. 1 mol/dm^3; 5. 0.5 mol/dm^3.

Sample IGCSE questions

1. Ammonium sulphate is made from a solution of ammonia, NH_3, and sulphuric acid, H_2SO_4.

(a) Write a balanced symbol equation for this reaction. **[3]**

$$2NH_3 + H_2SO_4 \rightarrow (NH_4)_2SO_4$$ ✓✓✓

There is a mark for the correct reactants, a mark for the correct products and a mark for balancing the equation correctly.

(b) Calculate the relative molecular mass of ammonium sulphate. **[2]**

$A_r(N) = 14$, $A_r(H) = 1$, $A_r(S) = 32$, $A_r(O) = 16$

Relative formula mass = $(2\times14)+(8\times1)+32+(4\times16)$ ✓
= 132 ✓

(c) Calculate the percentage of nitrogen in ammonium sulphate. **[3]**

Percentage of nitrogen = $\frac{(2 \times 14) \times 100}{132}$ ✓✓
= 21.2 ✓

There is one mark for multiplying 2 x 14, one for the expression and one for the correct answer.

(d) Explain why a farmer puts ammonium sulphate on a field and why it should not be done when heavy rain is forecast. **[3]**

Nitrogen helps the growth of stems and leaves ✓. Heavy rain may wash the ammonium sulphate out of the soil ✓. The fertiliser will be less effective or may cause water pollution problems ✓.

2. A sample of copper bromide, CuBr, weighing 21.6 g was heated with excess iron powder. A reaction took place producing copper and iron(III) bromide.

(a) Write a balanced symbol equation for the reaction. **[3]**

$$3CuBr + Fe \rightarrow 3Cu + FeBr_3$$ ✓✓✓

(b) What type of reaction took place? **[1]**

Displacement reaction ✓

(c) Calculate the mass of copper produced. $A_r(Cu) = 64$, $A_r(Br) = 80$ **[3]**

144 g of copper bromide ✓ produces 64 g of copper ✓
so 21.6 g of copper bromide produces 9.6 g of copper ✓

3. Magnesium reacts with hydrochloric acid according to the equation

$Mg(s) + 2HCl(aq) \rightarrow MgCl_2(aq) + H_2(g)$.

(a) What volume of hydrochloric acid (2 mol/dm^3) exactly reacts with 0.1 moles of magnesium atoms? **[4]**

Sample IGCSE questions

0.1 moles of magnesium atoms react with 0.2 moles of acid ✓. 1000 cm^3 of hydrochloric acid (2 mol/dm^3) contains 2 moles of hydrochloric acid ✓ 100 cm^3 of hydrochloric acid (2 mol/dm^3) contains 0.2 moles of hydrochloric acid ✓∴ 100 cm^3 of hydrochloric acid (2 mol/dm^3) reacts with 0.1 moles of magnesium atoms ✓

This comes from the balanced equation.

(b) What mass of hydrogen is produced when 0.1 moles of magnesium react with excess acid? **[4]**

1 mole of magnesium atoms reacts to produce 1 mole of hydrogen molecules ✓ 0.1 moles of magnesium atoms react to produce 0.1 moles of hydrogen molecules ✓ Mass of 1 mole of hydrogen molecules (H_2) = 2 g ✓
∴ Mass of hydrogen produced = 0.2 g ✓

Again it is important to use the balanced equation.

(c) What volume of hydrogen measured at room temperature and atmospheric pressure would be produced when 0.1 moles of magnesium atoms react with excess acid? **[4]**

1 mole of magnesium atoms reacts to produce 1 mole of hydrogen molecules ✓ 0.1 moles of magnesium atoms react to produce 0.1 moles of hydrogen molecules ✓ 1 mole of hydrogen molecules occupies 24 dm^3 at room temperature and atmospheric pressure ✓.
∴ Volume of hydrogen produced = $24 \times 0.1 = 2.4$ dm^3 ✓

In a question like this if you make an error in part (a) for example, the examiner will carry forward your answer to (b) and (c).

Exam practice questions

1. Andy carried out an experiment to check the concentration of some sulphuric acid being used in an electroplating process.

He titrated 25.0 cm^3 of the acid with sodium hydroxide solution containing 1.2 mol/dm^3 NaOH.

He found 35.0 cm^3 of the alkali was required for neutralisation.

The equation for the reaction is:

$2NaOH(aq) + H_2SO_4(aq) \rightarrow Na_2SO_4(aq) + 2H_2O(l)$

(a) What should he use to measure out **(i)** 25.0 cm^3 of acid; **(ii)** the alkali in small measured portions? **[2]**

(b) State, giving a named example, what he added in order to know when to stop adding the alkali. **[2]**

(c) **(i)** How many moles of sodium hydroxide (NaOH) were added in the titration? **[1]**

(ii) With how many moles of sulphuric acid, H_2SO_4, did 35.0 cm^3 of sodium hydroxide react? **[1]**

(iii) What is the concentration of the sulphuric acid in mol/dm^3? **[1]**

2. Vinegar contains ethanoic acid. An experiment was carried out to find the concentration of ethanoic acid in a sample of white vinegar.

50 cm^3 of vinegar was put into a flask. Sodium hydroxide solution (0.1 mol/dm^3) was added until the indicator changed colour. The volume of sodium hydroxide added was 20 cm^3.

The equation for the reaction is:

$CH_3COOH + NaOH \rightarrow CH_3COONa + H_2O$

(a) How many moles of sodium hydroxide are present in 20 cm^3 of 0.1 mol/dm^3 sodium hydroxide solution? **[2]**

(b) How many moles of ethanoic acid are present in 50 cm^3 of vinegar? **[1]**

(c) How many moles of ethanoic acid are present in 1000 cm^3 of vinegar? **[1]**

(d) What is the mass of 1 mole of ethanoic acid?

(Relative atomic masses $A_r(H) = 1$, $A_r(C) = 12$, $A_r(O) = 16$ **[1]**

(e) What is the concentration of ethanoic acid, in g/dm^3 in vinegar? **[2]**

Exam practice questions

3. The concept of the mole is used to calculate the amounts of chemicals involved in a reaction.

(a) Define mole. **[1]**

(b) 5.0 g of magnesium was added to 20.0 g of ethanoic acid.

$$Mg + 2CH_3COOH \rightarrow (CH_3COO)_2Mg + H_2$$

The mass of one mole of Mg is 24 g.

The mass of one mole of CH_3COOH is 60 g.

(i) Which one, magnesium or ethanoic acid, is in excess? Give reason. **[3]**

(ii) How many moles of hydrogen were formed? **[1]**

(iii) Calculate the volume of hydrogen formed, measured at r.t.p. **[2]**

(c) In an experiment, 30.0 cm of aqueous sodium hydroxide, 0.5 mol/dm^3, was neutralised by 25.0 cm of aqueous oxalic acid, $H_2C_2O_4$.

$$2NaOH + H_2C_2O_4 \rightarrow Na_2C_2O_4 + 2H_2O$$

Calculate the concentration of the oxalic acid in mol/dm^3.

(i) Calculate the number of moles of NaOH in 30.0 cm^3 of 0.5 mol/dm^3 solution. **[1]**

(ii) Use your answer to (i) and the mole ratio in the equation to find out the number of moles of $H_2C_2O_4$ in 25.0 cm of solution. **[1]**

(iii) Calculate the concentration, mol/dm^3, of the aqueous oxalic acid. **[2]**

Chapter 6 Air and water

The following topics will be covered in this section:

● Water ● Testing for water ● Purification of water supplies ● Oxygen ● Ammonia ● Changes in composition of atmosphere and oceans ● Air Pollution ● Carbon cycle ● Nitrogen fertilisers ● Rusting and its prevention

After studying this section you should be able to:

- ***describe the uses and tests for water***
- ***recall how a safe water supply can be produced***
- ***understand the process of manufacturing oxygen and enumerate its uses***
- ***recall that ammonia is made from nitrogen and hydrogen***
- ***understand the steps in producing ammonia by the Haber process***
- ***understand how the composition of the atmosphere has changed over history***
- ***name common air pollutants, their respective sources and effects***
- ***recall that a growing plant needs large quantities of nitrogen, phosphorus and potassium and reasons for the same***
- ***recall how one nitrogen fertiliser, ammonium nitrate, is made***
- ***understand the problems caused by the over-use of nitrogen fertilizers***
- ***recall that rusting of iron and steel requires both air and water to be present and describe ways of preventing rusting.***

Water

Water is a very important substance as it is essential for so many fundamental processes. It is, however, very difficult to get pure because it is very good at dissolving other substances. This ability to dissolve other things has important consequences.
A clean water supply is essential for good health. Many diseases e.g. cholera are a consequence of impure water supplies

Uses of water

Water is the foundation of life. It is the main constitutent of living matter. More than 50 percent of the weight of living organisms is water.

Domestic uses of water

- For drinking
- For cooking
- For washing and cleaning
- In air coolers

Industrial uses of water

- It is the most widely used coolant.
- Water is one of the best known ionising agents due to which it is commonly used in batteries.

- Being a very good solvent, it is used to dissolve impurites of materials insoluble in it.
- Water combines with certain salts to form hydrates.
- It reacts with metal oxides to form acids.
- It acts as a catalyst in many important chemical reactions.
- It is used for constructing and building structures.

Sea water contains a large amount of dissolved material. The most common ions in sea water are sodium ions and chloride ions.

When a sample of sea water is evaporated to half its original volume and left to cool, solid sodium chloride crystallises out.

Evaporation occurs at a surface and the larger the surface the faster the evaporation. Evaporation needs energy to occur.

The sea has been a traditional source of salt, which has been used for flavouring food but more importantly, before refrigeration, for preserving it.

In warm countries, large beds of sea water are left to evaporate using energy from the sun. These beds cover a large area and are shallow.

Testing for water

The presence of water in a liquid can be shown using anhydrous copper(II) sulphate or cobalt(II) chloride paper.

The results are shown in **Table 6.1** below:

Chemical used to detect water	Colour before testing	Colour after testing
Anhydrous copper(II) sulphate	*white*	*blue*
Cobalt(II) chloride paper	*blue*	*pink*

Candidates frequently get the colours wrong here especially as blue is the correct answer before with one chemical but after with the other.

A positive test with one of these chemicals shows that water is present: it does not show pure water. A solution of sodium chloride would give a positive test. To show that a liquid is pure water, melting and boiling point tests should be carried out.

Pure water freezes at 0°C and boils at 100°C.

Purification of water supplies

A clean, safe supply of water provided to homes and industry is essential for public health.

As a country becomes more developed, more water is required.

Every day each person in the United Kingdom uses about 120 litres of water – for washing, flushing toilets, cooking, etc. Industry too uses a lot of water – it takes 26 000 litres of water to make one tonne of newsprint, 45 500 litres to make a tonne of steel and up to 7 litres to make a pint of beer.

Producing tap water

Carbon in the form of charcoal is a good absorber of unwanted colours and tastes.

If the water contains iron compounds it can make tea have an inky colour and a bitter taste. It can also cause brown stains on clothes after washing.

There are various essential steps in producing tap water:

- water is taken from a clean river or reservoir
- the water is passed through a screen (sieve) to remove solid objects
- it is left to stand to allow solid material to settle out
- it is then filtered through a gravel bed to remove impurities in suspension
- it is then **chlorinated**. Chlorine is added in small amounts to kill bacteria and other harmful micro-organisms.

Other forms of treatment may be used in special cases.

These include:

Aluminium sulphate contains Al^{3+} which are very efficient at coagulation.

- adding aluminium sulphate to coagulate colloidal clay in water
- using a mixture of carbon and water to remove tastes and odours from the water
- using lime to correct acidity of the water
- adding sulphur dioxide to remove excess chlorine.

Fluoride is added to water to improve the dental health of consumers. Some people are against adding substances to water. Large concentrations of sodium fluoride are used as a rat poison.

The water produced at the end of all these processes is not pure water. It is water that is safe to drink.

Recycling waste water

The water leaving a sewage treatment plant is pure enough to be used again. Water in the River Thames is used several times for water supply before it enters the sea.

After water has been used by a home or a factory, it must be cleaned up before being returned to rivers and possibly re-used for water supply.

This is done in a sewage treatment plant. The treatment of waste water involves:

- filtering it to remove solid materials
- using bacteria to break down the waste.

PROGRESS CHECK

1. *Why is chlorine bubbled through water during treatment?*
2. *Which process is used to remove large objects such as leaves from water?*
3. *Which process removes suspended solids from water?*
4. *In the recycling of water, waste water is sprayed onto a large bed of gravel where much of the waste is broken down. What breaks down the waste?*
5. *Give one use of water in industry and at home.*
6. *Give two methods of testing purity of water.*

1. To kill harmful bacteria or other micro-organisms; 2. Sieving or screening;
3. Settling; 4. Bacteria; 5. Used as a coolant and for drinking/cooking;
6. Using cobalt chloride paper which turns pink and by checking B.P. or M.P.

Oxygen

Oxygen (symbol 'O') is a colourless, odourless and tasteless gaseous element. It is the most abundant element on Earth. Life on earth cannot survive without oxygen. The percentage of oxygen in air is 21% by volume. Oxygen has many commercial uses too for which it is extracted from air.

Manufacture and uses of oxygen

Oxygen is manufactured by fractional distillation of liquefied air. Air is first cleaned by filtering out dust, then cooled to solidify–this removes the water vapour and carbon content.

Then, it is compressed and cooled further and again allowed to expand which causes further cooling down to below (-)200 °C. At this temperature, most of the air is liquified. The liquid air is seperated by fractional distillation according to the following boiling points.

Nitrogen is the first to separate followed by argon, oxygen, krypton and xenon.

Name of Gas	Xenon(Xe)	Krypton(Kr)	Oxygen(O)	Argon(Ar)	Nitrogen(N)
Boiling Point (°C)	–108	–153	–183	–186	–196

Oxygen has a large number of uses, some of which are:

1. for patients in hospitals and for divers, firefighters, astronauts etc., oxygen is used for **breathing purposes**.
2. In **steel making**: Oxygen is blown through molten iron to oxidise all impurities and convert pig iron to steel.
3 In **oxy-acetylene cutting and welding** equipment: when burnt with gases such as acetylene, it produces very high temperatures, sufficient to melt metals like iron.
4. It is used in rockets to enable **burning of rocket fuel** in space.

Ammonia

Ammonia, NH_3, is a compound of nitrogen and hydrogen. It is produced in large quantities by the Haber process using nitrogen from the air as a raw material.

The process is summarised in **Fig. 6.1.**

It is not possible to get 100% conversion to products in a reversible reaction such as this.

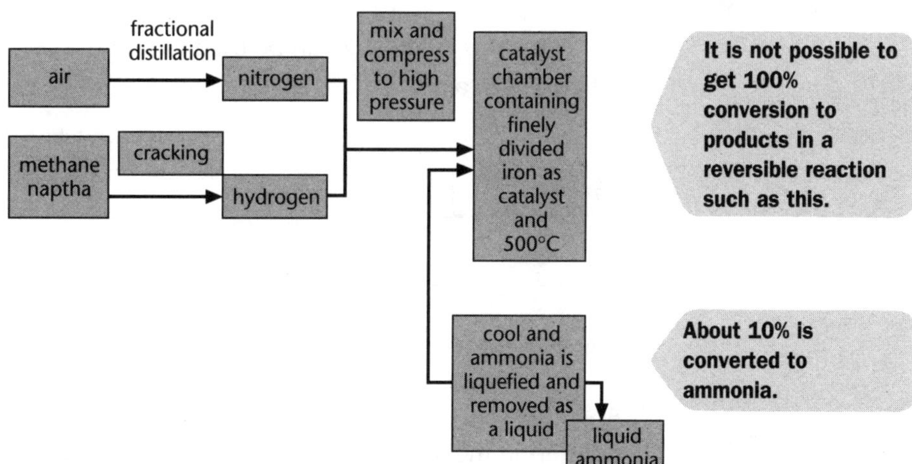

Unreacted nitrogen and hydrogen recycled.

About 10% is converted to ammonia.

Fig. 6.1 Haber process

The equation for the reaction is:

$$N_2 + 3H_2 \rightleftharpoons 2NH_3$$

The usual arrow between the reactants and the products is replaced with a **reversible reaction sign**. This means that the products can decompose, reforming the reactants.

By choosing the best conditions, chemists attempt to produce the highest **yield** of ammonia economically.

The best conditions are:

Catalysts are usually transition metals or transition metal compounds.

1. **one part of nitrogen to 3 parts of hydrogen by volume**
2. **a high pressure**
3. **a low temperature**. However, using a low temperature reduces the rate of reaction. Using an **iron catalyst** speeds up the reaction.

In practice, the Haber process operates at a temperature of about 450°C and is used with a catalyst of iron. A high pressure, e.g. 200 atmospheres, is used.

The process is called the **Haber process** after the German scientist, *Fritz Haber*, who discovered the conditions which would enable ammonia to be made on a large scale.

Changes in composition of atmosphere and oceans

Composition of the present atmosphere

Air is a **mixture** of gases. Its composition can vary from place to place.

The typical composition of a sample of dry air is

Nitrogen	78%
Oxygen	21%
Argon (and other noble gases)	1%
Carbon dioxide	0.04%

Students frequently write that air contains hydrogen. Normal air does not.

The apparatus in **Fig. 6.2** can be used to find the percentage of oxygen (the active gas) in air.

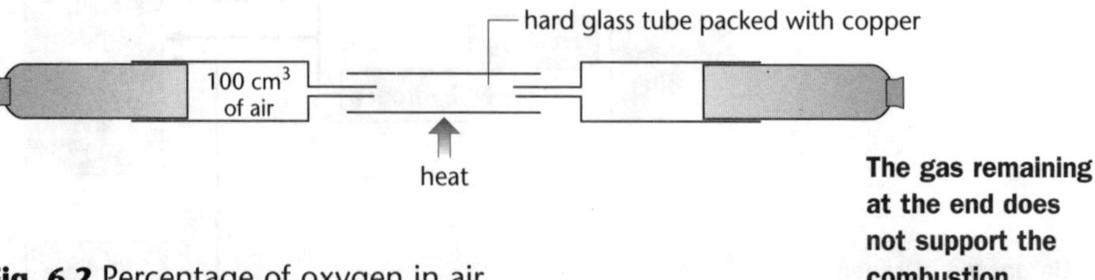

The gas remaining at the end does not support the combustion.

Fig. 6.2 Percentage of oxygen in air

It is important that all volumes are measured at room temperature. At higher temperature gases would be expanded.

A sample of air is passed backwards and forwards over heated **copper**. The **oxygen** in the air is removed. Black **copper(II) oxide** is formed.

$$2Cu + O_2 \rightarrow 2\ CuO$$

The percentage of oxygen in the air can be calculated by measuring the remaining volume after the apparatus has cooled to room temperature.

How the atmosphere has changed

Table 6.2 Effects of changes on the atmosphere

Change	Effect on the atmosphere
The first atmosphere	*Consisted mainly of* **hydrogen** *and* **helium**
Volcanoes started to erupt	*Mostly* **carbon dioxide** *and* **water vapour** *entering the atmosphere. Smaller quantities of* **methane** *and* **ammonia**
Earth cools	*Water vapour condenses to* **liquid water**. *Oceans started to form*
Nitrifying and denitrifying bacteria start to work	*Ammonia is converted into* **nitrates,** *and* **nitrates** *are converted into gaseous* **nitrogen**
Methane in the atmosphere burns	**Carbon dioxide** *is formed*
Photosynthesis occurs	*Plants convert carbon dioxide into* **oxygen**
Increasing levels of	*Due to burning of fossil fuels and destruction*

Air pollution

In the past, changes in atmosphere ocurred mainly due to natural causes.

In recent times, due to rapid growth in population and consequent industrialisation, the air is getting increasingly polluted.

KEY POINT **Most of the air pollution nowadays is caused by burning of fossil fuels like coal, oil and gas**

Sulphur dioxide and oxides of nitrogen are ultimately converted to sulphuric acid and nitric acid respectively, in the atmosphere, which leads to acid rain.

Table 6.3 Main pollutants of air

Main pollutants of air	Formula	Main source	Fossil fuel used	Cause	Adverse effects
Carbon monoxide	*CO*	*Car exhausts*	*Oil(Petrol/ Diesel)*	*Inefficient combustion*	*Toxic gas, can cause death.*
Sulphur dioxide	SO_2	*Power stations*	*Coal and gas*	*Burning of sulphur present in fossil fuels*	*Main cause of acid rain. A lung irritant- causes respiratory diseases. Leads to corrosion of building stones (particularly limestone).*
Hydrocarbons	C_xH_y	*Engine exhaust*	*Oil(Petrol/ Diesel)*	*Inefficient combustion*	*Carcinogenic- increases risk of cancer, contributes to 'photochemical smog'.*

(continued...)

Table 6.3 *(continued...)*

Oxides of Nitrogen	*NO and NO_2*	*Power stations and engines*	*Coal, oil and gas*	*Burning of nitrogen present in fuels*	*Leads to 'photochemical smog'- causes respiratory diseases, eye irritation and contributes to acid rain.*
Lead compounds	*$PbNO_3$ etc.*	*Car exhausts*	*Oil(Petrol/ Diesel)*	*Fuel additives*	*Nerve toxins, causes damage to the brain and nervous system*

In car engines, nitrogen present in petrol/diesel reacts with oxygen to form nitrogen.

$$\text{Nitrogen} + \text{Oxygen} \rightarrow \text{nitrogen monoxide}$$
$$N_2(g) + O_2(g) \rightarrow 2NO(g)$$

When emitted from car exhausts, nitrogen monoxide rapidly combines with oxygen in the air.

$$\text{Nitrogen monoxide} + \text{Oxygen} \rightarrow \text{Nitrogen dioxide (acidic gas)}$$
$$2NO(g) + O_2(g) \rightarrow 2NO_2(g)$$

The nitrogen dioxide is oxidised to nitric acid by the reaction with oxygen from air, when it dissolves in rainwater. This causes acid rain.

The overall process is summarised in the equation below.

$$4NO_2(\text{g-air}) + O_2(\text{g-air}) + 2H_2O(\text{l-rain}) \rightarrow 4HNO_3(\text{aq-rain})$$

Air pollution from automobiles (e.g. cars) can be reduced by:

- use of **unleaded petrol**
- use of **catalytic convertors**
- use of engines with higher **fuel efficiency**
- use of gases like CNG and LPG as fuels instead of petrol and diesel
- use of battery operated engines
- use of **alternative energy resources**, e.g. solar cells.

Catalytic converters are used to reduce the quantity of carbon monoxide and nitrogen monoxide, emitted from car exhausts, by converting the two gases into harmless nitrogen and carbon dioxide.

$$2NO(g) + 2CO(g) \rightarrow N_2(g) + 2CO_2(g)$$

Carbon cycle

KEY POINT **The concentration of carbon dioxide has been kept constant by a delicate balance between respiration and combustion on one side and photosynthesis on the other.**

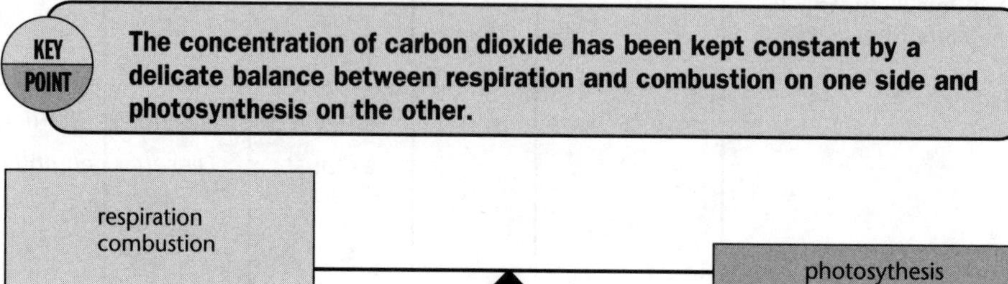

Fig. 6.3 Balance between respiration, combustion and photosynthesis

Over the last 50 years, this balance has been disturbed. The main reason for this is the **increased burning of fossil fuels** producing carbon dioxide and the destruction of forests and green plants which would remove carbon dioxide.

The increasing **concentration of carbon dioxide** (and other greenhouse gases) has caused a rise in the average temperature of the Earth's atmosphere called **global warming**.

Fig. 6.4 shows how the **greenhouse effect** brings about global warming.

One hectare (2.47 acres) of Sitka Spruce trees in a year uses up 6 tonnes of carbon dioxide, producing four tonnes of oxygen.

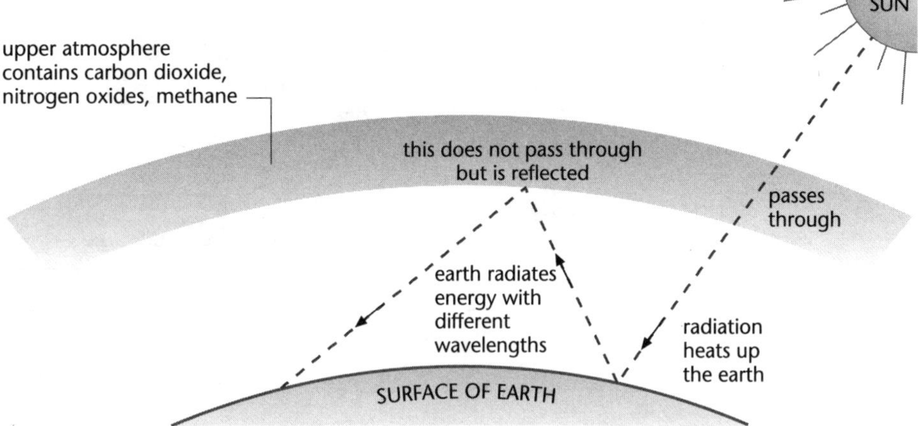

Fig. 6.4 Greenhouse effect

1. *Which gas is the commonest gas in the atmosphere?*
2. *Which compound is the most common in the atmosphere?*
3. *Which gas in the atmosphere is the active gas?*
4. *Which gas makes up about one-fifth of the atmosphere?*
5. *Which process removes carbon dioxide from the atmosphere?*
6. *Which processes produce carbon dioxide?*

1. Nitrogen; 2. Carbon dioxide; 3. Oxygen; 4. Oxygen; 5. Photosynthesis;
6. Combustion (of fossil fuels) and respiration.

Nitrogen fertilisers

Three elements, **nitrogen**, **phosphorus** and **potassium** are required in large quantities by a healthy plant.

Table 6.4 summarises the importance of these elements to the growing plant and gives some natural and artificial sources.

Table 6.4 Source and use of elements in fertilisers

Element	Importance to growing plant	Natural sources	Artificial sources
Nitrogen	*For growth of stems and leaves*	*Manure, bird droppings, dried blood*	*Ammonium nitrate, ammonium sulphate, urea*
Phosphorus	*For root growth*	*Bone meal*	*Ammonium phosphate*
Potassium	*For flowers and fruit*	*Wood ash*	*Potassium sulphate*

In the past 100 years, there has been a **huge growth in population** and therefore in the **quantity of food** required to feed everybody. The development of better and cheaper fertilisers has enabled food production to increase.

Fig. 6.5 summarises how one fertiliser, ammonium nitrate, is made from ammonia produced in the Haber process.

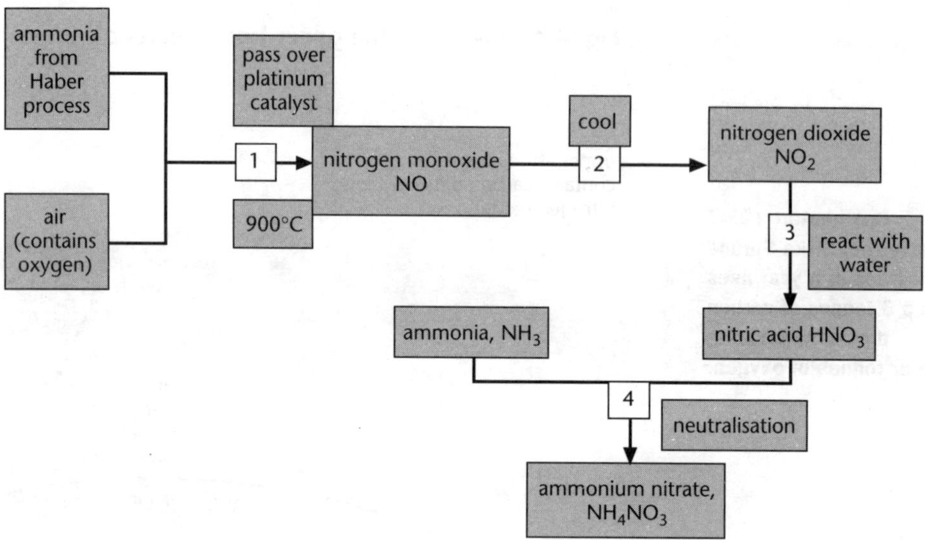

Fig. 6.5 Ammonium nitrate production

The equations for the reactions taking place are:

This is one of the hardest equations to balance that you will find.

1. ammonia + oxygen → nitrogen monoxide + water

   $4NH_3(g) + 5O_2(g) \rightarrow 4NO(g) + 6H_2O(g)$

2. nitrogen monoxide + oxygen ⇌ nitrogen dioxide

   $2NO(g) + O_2(g) \rightleftharpoons 2NO_2(g)$

3. nitrogen dioxide + water + oxygen → nitric acid

   $4NO_2(g) + 2H_2O(l) + O_2(g) \rightarrow 4HNO_3(l)$

4. nitric acid + ammonia → ammonium nitrate

   $HNO_3(aq) + NH_3(aq) \rightarrow NH_4NO_3(aq)$

These equations refer to the steps in the flow diagram of Fig. 6.5.

When ammonium salts come in contact with a more reactive compound, the ammonium ion is **displaced** by the more reactive ion and ammonia gas is released as a result.

E.g.

Ammonium Chloride + Sodium Hydroxide → Sodium Chloride + Ammonia + Water

$NH_4Cl(aq) + NaOH(aq) \rightarrow NaCl(aq) + NH_3(g) + H_2O(l)$

Here, the more reactive Na^+ ion replaces the less reactive NH_4^+ ions.

Over-use of nitrogen fertilisers

A similar result is encountered, if sewage waste escapes into a river.

Nitrogen fertilisers in the soil are turned into **nitrates**. These are absorbed into plants in solution through the **roots**.

Nitrates are very soluble and so can be washed out of the soil by rain. If they are washed into rivers a series of changes may take place. This leads to **eutrophication** when there is little life left in the river.

1. Nitrates make water plants grow and these cover the surface of the river.
2. These shade the surface, preventing **light** getting into the water and stopping **photosynthesis**.
3. When these plants die, **bacteria** in the river decompose them.
4. These bacteria use up **oxygen**.
5. There is little oxygen left dissolved in the water, and fish and other life die.

Nitrates in water can cause problems to health e.g. blue baby syndrome.

PROGRESS CHECK

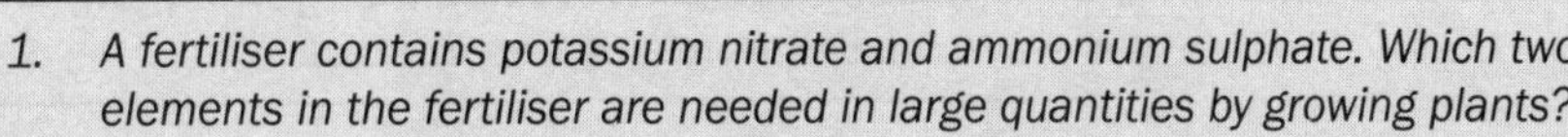

1. *A fertiliser contains potassium nitrate and ammonium sulphate. Which two elements in the fertiliser are needed in large quantities by growing plants?*
2. *Ammonia is a compound of two elements? What are these two elements?*
3. *What is the meaning of the sign $\rightleftharpoons$ in the reaction?*
4. *What is the catalyst in the Haber process?*
5. *What is the source of nitrogen and hydrogen in the Haber process?*
6. *Ammonium nitrate dissolves readily in water. Urea does not dissolve in water but reacts with water slowly to produce ammonia. Suggest when each fertiliser would be used.*

1. Nitrogen and potassium; 2. Nitrogen and hydrogen; 3. Reversible reaction; 4. Iron; 5. nitrogen from the air, hydrogen from methane or naphtha; 6. Ammonium nitrate is a quick-acting fertiliser and urea is a slow-acting fertiliser.

Rusting and its prevention

Iron and steel react in the atmosphere to produce reddish-brown rust.

The chemical composition of rust is complicated. It is best regarded as a hydrated iron(III) oxide, $Fe_2O_3.xH_2O$. The rusting process is an **oxidation** reaction and can be represented by the equation

$$Fe \rightarrow Fe^{3+} + 3e^-$$

A surface coating of aluminium oxide forms on a piece of aluminium and prevents further reaction. Rust, however, flakes off and shows a fresh metal surface for rusting.

Fig. 6.6 below shows an experiment used to find the conditions needed for rusting.

Fig. 6.6

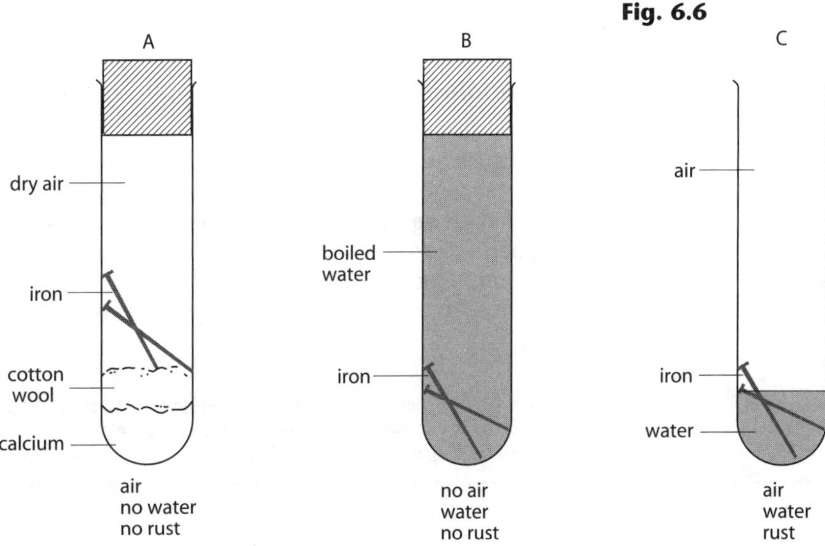

If the three test tubes are set up as shown in **Fig. 6.6** and left for a few days, rusting only occurs in test tube C.

From this it can be concluded that:

KEY POINT **Both air and water are needed for rusting to take place.**

In fact, it is the **oxygen** in the air along with the **water** that are needed for rusting.

Rusting is speeded up when acid (carbon dioxide or sulphur dioxide) or salt are present.

Ways of preventing rusting

KEY POINT **Rusting can be prevented if air (or oxygen) and/or water can be excluded from the iron or steel.**

1. **Painting**. A coat of paint prevents oxygen and water from coming into contact with the iron or steel. This will prevent rusting only while the coating is intact. This type of rust prevention is used to prevent steel car bodies or iron railings from rusting.
2. **Oil or grease**. These prevent oxygen and water from coming in contact with iron and steel. They are useful for moving parts.
3. **Coating with zinc**. A layer of zinc on iron or steel prevents rusting by stopping oxygen and water from coming into contact with the metal. This is called **galvanising**. It is used for fences. It cannot be used for food cans as zinc compounds are poisonous. Tin plate is used to make food cans.
4. **Sacrificial protection**. This is used for preventing the rusting of steel hulls of ships or steel legs of piers. To prevent the rusting of a steel hull blocks of **magnesium** or **zinc** are strapped to it. The metal used must be higher in the reactivity series. The magnesium or zinc blocks corrode in preference to iron. As long as they remain, no rusting will take place. These blocks can easily be replaced.
5. **Electroplating**. A steel item can be protected from rusting by electroplating. A thin layer of **nickel** is plated on the surface to prevent oxygen and water from coming in contact with the steel. A very thin coating of **chromium** is often plated on top to give a decorative appearance.

PROGRESS CHECK

1. *What substances are needed for rusting of iron and steel to take place?*
2. *Write the ionic equation for the rusting of iron.*
3. *Which of these metals in contact with iron and steel will slow down the rusting?*
   copper lead magnesium zinc
4. *Write down the formula of rust.*
5. *Why does a car exhaust made of steel rust faster than other steel parts of the car?*

1. Air (or oxygen) and water; 2. $Fe \rightarrow Fe^{3+} + 3e^-$; 3. Magnesium and zinc (above iron in the reactivity series); 4. $Fe_2O_3.xH_2O$; 5. Exhaust is hot so rusting is speeded up; acidic gases are inside the exhaust; it is impossible to paint; it is closer to the surface of road so picks up more salt and water from the road.

Sample IGCSE questions

1. State two conditions for the Haber process. **[2]**

Temperature about 450 °C, high pressure, iron catalyst ✓✓

Any two conditions required. There are 3 in this answer.

2. Sam finds a colourless liquid. She wants to find out if this liquid contains water.

She adds anhydrous copper(II) sulphate to the liquid.

(a) What colour change would she expect if the liquid contains water? **[2]**

White ✓ to blue ✓

When talking about a change, you need to give the initial and final colours.

She adds Universal Indicator to the liquid and it turns green (pH 7). She evaporates a sample to dryness and a white residue is obtained. The liquid boils at 102°C.

(b) Which of her observations suggest that the liquid could be pure water and which suggest the liquid could be an aqueous solution? **[2]**

Boils at 102 °C or white residue of evaporation–aqueous solution. ✓ pH 7–pure water ✓

Exam practice questions

1. Ammonia contains the elements nitrogen and hydrogen. It is manufactured from these elements in an exothermic process called the **Haber process**.

$$N_2(g) + 3H_2(g) \rightleftharpoons 2NH_3(g)$$

(a) **(i)** Nitrogen is obtained from liquid air by fractional distillation. How does this technique separate liquid oxygen and nitrogen?

(ii) How is hydrogen for Haber's process, manufactured? **[3]**

(b) The table shows the variation of percentage of ammonia in the equilibrium mixture with pressure at 600 °C.

percentage ammonia	9	13	15	22
pressure/atm	250	350	450	550

(i) Explain why the percentage of ammonia increases as the pressure increases. **[2]**

(ii) How would the percentage of ammonia change, if the measurements had been made at a lower temperature? Explain your answer. **[2]**

(iii) State **two** of the reaction conditions used in the Haber process. **[2]**

(c) Ammonia is a base.

(i) Name a particle that an ammonia molecule can accept from an acid.

(ii) Write an equation for ammonia acting as a base. **[3]**

Chapter

7 The periodic table

The following topics will be covered in this section:

- ***Periodic trends***
- ***Group properties***
- ***Noble gases***
- ***Transition metals***

7.1 Periodic trends

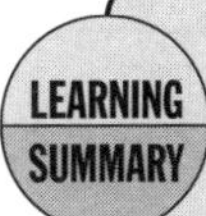

After studying this section you should be able to:

- ***describe the structures of atoms of the first 20 elements***
- ***explain the link between reactivity and electron arrangement.***

Arrangement of electrons in an atom

The electrons move rapidly around the nucleus in distinct energy levels. Each energy level is capable of holding only a certain maximum number of electrons. This is represented in a simplified form in **Fig. 7.1**.

Beyond element 20 (calcium), the order of filling energy levels is slightly different.

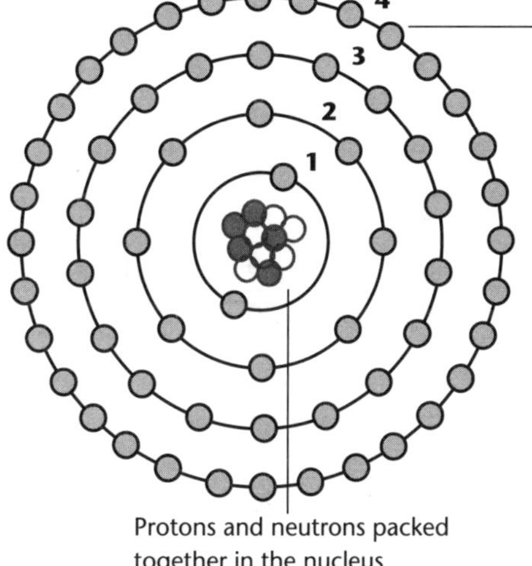

Fig. 7.1 Arrangement of particles in an atom

These energy levels are sometimes called 'shells'.

- The **first energy level** (labelled 1 in **Fig. 7.1**) can hold only **two electrons**. This energy level is filled first.
- The **second energy** level (labelled 2 in **Fig. 7.1**) can hold only **eight electrons**. This energy level is filled after the first energy level and before the third energy level.
- The **third energy** level (labelled 3 in **Fig. 7.1**) can hold a maximum of **18 electrons**. However, when eight electrons are in the third energy level there is a degree of stability and the next two electrons added go into the fourth energy level (labelled 4 in **Fig. 7.1**). Then extra electrons enter the third energy level until it contains the maximum of 18 electrons.
- There are further energy levels, each containing a larger number of electrons than the preceding energy level.

Table 7.1 gives the number of protons, neutrons and electrons in the principal isotopes of the first 20 elements. The arrangement of electrons 2,8,1 denotes 2 electrons in the first energy level, 8 in the second, and 1 in the third. This is sometimes called the **electron arrangement** or **electronic configuration** of an atom.

You ought to be able to draw simple diagrams of atoms of the first 20 elements. Don't forget to show the nucleus and all energy levels.

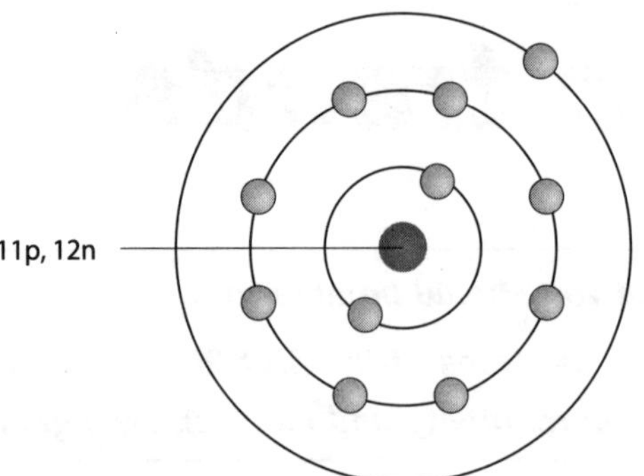

Atoms are sometimes shown in simple diagrams. **Fig. 7.2** shows a sodium atom.

Fig 7.2 A sodium atom

Do not try to remember all of the information in the table. You will be able to work it out from the periodic table.

The noble gases* have their outermost shells completely filled with electrons. e.g. Helium-2, Neon-8, Argon-8 etc.

Table 7.1 Numbers of protons, neutrons and electrons in the principal isotopes of the first 20 elements.

Element	Atomic number	Mass number	Number of p	n	e	Electron arrangement
Hydrogen	*1*	*1*	*1*	*0*	*1*	*1*
*Helium**	*2*	*4*	*2*	*2*	*2*	*2*
Lithium	*3*	*7*	*3*	*4*	*3*	*2,1*
Beryllium	*4*	*9*	*4*	*5*	*4*	*2,2*
Boron	*5*	*11*	*5*	*6*	*5*	*2,3*
Carbon	*6*	*12*	*6*	*6*	*6*	*2,4*
Nitrogen	*7*	*14*	*7*	*7*	*7*	*2,5*
Oxygen	*8*	*16*	*8*	*8*	*8*	*2,6*
Fluorine	*9*	*19*	*9*	*10*	*9*	*2,7*
*Neon**	*10*	*20*	*10*	*10*	*10*	*2,8*
Sodium	*11*	*23*	*11*	*12*	*11*	*2,8,1*
Magnesium	*12*	*24*	*12*	*12*	*12*	*2,8,2*
Aluminium	*13*	*27*	*13*	*14*	*13*	*2,8,3*
Silicon	*14*	*28*	*14*	*14*	*14*	*2,8,4*
Phosphorus	*15*	*31*	*15*	*16*	*15*	*2,8,5*
Sulphur	*16*	*32*	*16*	*16*	*16*	*2,8,6*
Chlorine	*17*	*35*	*17*	*18*	*17*	*2,8,7*
*Argon**	*18*	*40*	*18*	*22*	*18*	*2,8,8*
Potassium	*19*	*39*	*19*	*20*	*19*	*2.8.8.1*
Calcium	*20*	*40*	*20*	*20*	*20*	*2,8,8,2*

Link between reactivity and electron arrangement

KEY POINT **The reactivity of elements is related to the electron arrangement in their atoms.**

Elements with atoms having **full electron energy** levels are very **unreactive**. These electron arrangements are said to be **stable**. It was believed at one time that they never reacted. These elements include helium, neon and argon.

Reactive elements have atoms containing nearly empty or nearly full outer energy levels.

Elements with atoms containing **one or two electrons** in the outer energy level are very **reactive**. These atoms tend to lose these outer electrons and finish up with a stable electron arrangement.

Elements with atoms containing **six or seven electrons** in the outer energy level are also **very reactive**. These atoms tend to gain one or more extra electrons and finish up with, again, a stable electron arrangement.

Since the combining power or valency of atoms is related to the number of outer electrons, therefore these electrons are called valence electrons.

Elements with atoms containing three, four or five electrons in the outer energy level are usually less reactive.

Table 7.2 gives some reactive and some unreactive elements. It also gives the arrangement of electrons in atoms of these elements.

Hydrogen is interesting. It has an electron arrangement of 1. It is reactive but it can gain 1 electron or lose 1 electron.

Reactive elements		**Unreactive elements**	
oxygen	*2, 6*	*carbon*	*2, 4*
chlorine	*2, 8, 7*	*silicon*	*2, 8, 4*
fluorine	*2, 7*	*nitrogen*	*2, 5*
		boron	*2, 3*
sodium	*2, 8, 1*		
potassium	*2, 8, 8, 1*		
calcium	*2, 8, 8, 2*		

In the reactive elements column, the horizontal line separates elements that are reactive because they gain electrons (above the line) from those that are reactive because they lose electrons. From the arrangement of electrons, you can make a prediction whether an element is reactive or unreactive.

PROGRESS CHECK

1. *Which particles are always present in equal numbers in an atom?*
2. *Which particles are in the nucleus of an atom?*
3. *Iron has an atomic number of 26 and a mass number of 56.*
   What are the numbers of protons, neutrons and electrons in an iron atom?
4. *There are three isotopes of hydrogen: Hydrogen-1, Hydrogen-2, Hydrogen-3.*
   How are atoms of these three isotopes different?
   Refer back to Table 7.1.
   Which of these statements are true and which are false?
5. *The elements are arranged in order of increasing atomic number.*
6. *The number of protons and neutrons is always the same.*
7. *The number of neutrons is always equal to or greater than the number of protons.*
8. *The number of neutrons is usually but not always even.*
9. *Which atom has two filled energy levels?*
10. *Which atom is shown in the diagram below?*

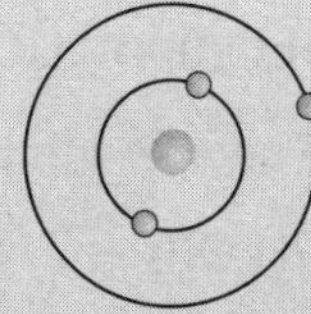

1. Protons and electrons; 2. Protons and neutrons; 3. 26p, 26e, 30n; 4. Different numbers of neutrons: Hydrogen-1 no neutrons, Hydrogen-2 one neutron, Hydrogen-3 two neutrons; 5. True; 6. False; 7. False (look at hydrogen); 8. True; 9. Neon; 10. Lithium.

7.2 Group Properties

LEARNING SUMMARY

After studying this section you should be able to:

- ***understand the patterns within families of elements: alkali metals and halogens***
- ***compare the physical properties of alkaline earth metals and see trends within the family***
- ***describe the trends in chemical reactions with oxygen and water for Group 2 metals.***

Properties and reactions of alkali metals

The alkali metals are a family of **very reactive metals**. The most common members of the family are lithium ,sodium and potassium. Some of the properties of these elements are shown in **Table 7.3** below.

Element	Symbol	Appearance	Melting point in °C	Density In g/cm³
Lithium	*Li*	*Soft grey metal*	*181*	*0.54*
Sodium	*Na*	*Soft light grey metal*	*98*	*0.97*
Potassium	*K*	*Very soft blue/grey metal*	*63*	*0.86*

These metals have to be stored in oil to exclude air and water. They do not behave much like metals, at first sight, but when **freshly cut they all have a typical shiny metallic surface**.

They are also **very good conductors of electricity**. Note, however, that they have **melting points and densities that are low** compared with other metals.

In the periodic table the alkali metals are in group 1.

Reaction of alkali metals with water

When a small piece of an alkali metal is put into a trough of water, the metal reacts immediately, floating on the surface of the water and evolving **hydrogen**.

With sodium and potassium, the heat evolved from the reaction is sufficient to melt the metal.

The hydrogen evolved by the reaction of potassium with cold water is usually ignited and burns with a pink flame.

Sodium reacts more quickly than lithium, and potassium reacts more quickly than sodium.

In each case the solution remaining at the end of the reaction is an alkali.

$2Li(s) + 2H_2O(l) \rightarrow 2LiOH(aq) + H_2(g)$

lithium + water → lithium hydroxide + hydrogen

$2Na(s) + 2H_2O(l) \rightarrow 2NaOH(aq) + H_2(g)$

sodium + water → sodium hydroxide + hydrogen

$2K(s) + 2H_2O(l) \rightarrow 2KOH(aq) + H_2(g)$

potassium + water → potassium hydroxide + hydrogen

The last three equations are basically the same and, if the alkali metal is represented by M, these equations can be represented by:

$$\mathbf{2M(s) + 2H_2O(l) \rightarrow 2MOH(aq) + H_2(g)}$$

Reaction of alkali metals with oxygen

When heated in air or oxygen, the alkali metals burn to form white solid **oxides**. The colour of the flame is characteristic of the metal:

lithium — red

sodium — orange

potassium — lilac

E.g. $4Li(s) + O_2(g) \rightarrow 2Li_2O(s)$

lithium + oxygen → lithium oxide

or $4M(s) + O_2(g) \rightarrow 2M_2O(s)$

The **alkali metal oxides** all dissolve in water to form **alkali solutions**.

E.g. $Li_2O(s) + H_2O(l) \rightarrow 2LiOH(aq)$

lithium oxide + water → lithium hydroxide

or $M_2O(s) + H_2O(l) \rightarrow 2MOH(aq)$

Reaction of alkali metals with chlorine

When a piece of burning alkali metal is lowered into a gas jar of chlorine, the metal continues to burn forming a white smoke of the metal **chloride**.

E.g. $2K(s) + Cl_2(g) \rightarrow 2KCl(s)$

potassium + chlorine → potassium chloride

or $2M(s) + Cl_2(g) \rightarrow 2MCl(s)$

It is because of these similar reactions that these metals are put in the same family. In each reaction, the order of reactivity is the same, i.e., **lithium is the least reactive and potassium is the most reactive**.

There are three more members of this family: rubidium (Rb), caesium (Cs) and francium (Fr). All three of them are more reactive than potassium.

Properties and reactions of halogens

The halogens are a family of non-metals.

In the halogen family, the different elements have different appearances but they are in the same family on the basis of their similar chemical properties.

Table 7.4 compares the appearances of four of these elements.

Candidates frequently spell fluorine as flourine.

Element	Symbol	Appearance at room temperature
Fluorine	*F*	*Pale yellow gas*
Chlorine	*Cl*	*Yellow/green gas*
Bromine	*Br*	*Red/brown volatile liquid*
Iodine	*I*	*Dark grey crystalline solid*

> In the periodic table the halogens are in group 7.

There is another member of the family called astatine (At). It is radioactive and a very rare element.

Fluorine is a very reactive gas and is too reactive to handle in normal laboratory conditions.

Solubility of halogens in water

None of the halogens is very soluble in water. Chlorine is the most soluble. Iodine does not dissolve much in cold water and only dissolves slightly in hot water.

> Halogens contain molecules with covalent bonding. They dissolve better in organic solvents e.g. hexane.

Chlorine solution (sometimes called chlorine water) is very pale green. It turns Universal Indicator red, showing that the solution is **acidic**. The colour of the indicator is quickly bleached.

Bromine solution (bromine water) is orange. It is very weakly acidic and also acts as a bleach.

Iodine solution is very weakly acidic and is also a slight bleach. The low solubility of halogens in water (a polar solvent) is expected because halogens are composed of molecules.

Solubility of halogens in hexane (a non-polar solvent)

The halogens dissolve readily in hexane to give solutions of characteristic colour:

chlorine – colourless

bromine – orange

iodine – purple.

Reactions of halogens with iron

The halogens react with **metals** by direct combination to form **salts**. The name **'halogen' means salt producer. Chlorine** forms **chlorides, bromine** forms **bromides and iodine** forms **iodides.**

If **chlorine** gas is passed over heated **iron** wire, an exothermic reaction takes place forming **iron(III) chloride**, which forms as a brown solid on cooling.

Fig. 7.3 shows a suitable apparatus for preparing anhydrous iron(III) chloride crystals.

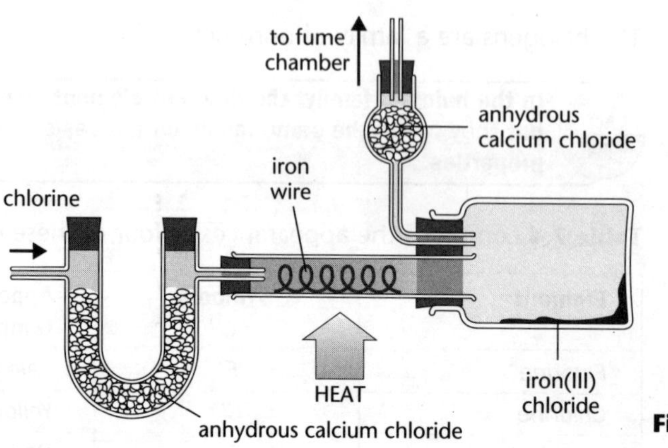

Fig. 7.3

$2Fe(s) + 3Cl_2(g) \rightarrow 2FeCl_3(s)$

iron + chlorine → iron(III) chloride

Bromine vapour also reacts with hot iron wire to form iron(III) bromide. When iodine crystals are heated, they turn to a purple vapour. This vapour reacts with hot iron wire to produce iron(II) iodide.

Order of reactivity of the halogens

From their chemical reactions the relative reactivities of the halogens are:

The reactivity of halogens decreases down the group.

fluorine	**most reactive**
chlorine	↓
bromine	↓
iodine	**least reactive**

Displacement reactions of the halogens

A more reactive halogen will displace a less reactive halogen from one of its compounds.

For example, when **chlorine** is bubbled into a solution of **potassium bromide**, the chlorine displaces the less reactive bromine. This means the colourless solution turns orange as free bromine is formed.

$2KBr(aq) + Cl_2(g) \rightarrow 2KCl(aq) + Br_2(aq)$

potassium bromide + chlorine → potassium chloride + bromine

No reaction would take place if iodine solution were added to potassium bromide solution because iodine is less reactive than bromine.

Properties of alkaline earth metals

The alkaline earth metals are all metals. They conduct electricity.

You know the trend in properties of the metals in group 1 of the periodic table – lithium, sodium, potassium, rubidium and caesium. There are similar trends in properties the elements in group 2 of the periodic table. These elements, called the alkaline earth metals, are beryllium, magnesium, calcium, strontium and barium.

Table 7.5 below gives the melting points, boiling points and densities of elements in group 2 of the periodic table.

Element	Melting point (°C)	Boiling point (°C)	Density (g/cm³)
Beryllium	*1278*	*2970*	*1.85*
Magnesium			*1.74*
Calcium	*839*	*1484*	*1.54*
Strontium	*769*	*1384*	*2.60*
Barium	*725*	*1640*	*3.51*

Melting and boiling points of group 2 elements are much higher than those of group 1.

There is a trend in the melting and boiling points of the metals in group 2 of the periodic table.

KEY POINT
The melting and boiling points of the elements decrease down the group.

There is an exception in the trends. Magnesium seems to have a lower melting point and boiling point than you would expect from the melting and boiling points of the other elements.

KEY POINT
The densities of alkaline earth metals are greater than the densities of alkali metals. None float on water.

1. *Which of the alkaline earth metals has the highest melting and boiling point?*
2. *Using the data in the table, what would you expect the melting point of magnesium to be?*
3. *Using the data in the table, what would you expect the boiling point of magnesium to be?*
4. *Radium is below barium in group 2. Suggest a melting and boiling point for radium.*

1. Beryllium; 2. Around 1000°C; 3. Around 2000°C; 4. Around 700°C and around 1200°C.

Reactions with air and water

Magnesium also reacts with nitrogen to form magnesium nitride. Lithium behaves in a similar way.

The alkaline earth metals are less reactive than the alkali metals in group 1 of the periodic table.

There is a trend in the reactivity of the elements.

Reactions with air

Magnesium burns brightly in air or oxygen to form **magnesium oxide**.

$2Mg(s) + O_2(g) \rightarrow 2MgO(s)$

Magnesium oxide is a white solid. A damp piece of universal indicator paper turns purple, showing the formation of an alkaline oxide.

Calcium burns in air or oxygen to form **calcium oxide**.

$2Ca(s) + O_2(g) \rightarrow 2CaO(s)$

Reactions with water

Magnesium hydroxide would be formed but at the high temperature it decomposes into magnesium oxide and steam.

Magnesium hardly reacts with cold water and reacts very slowly with boiling water.

It does react rapidly with steam to produce **magnesium oxide** and **hydrogen**.

$Mg(s) + H_2O(g) \rightarrow MgO(s) + H_2(g)$

Calcium reacts slowly with cold water to form **calcium hydroxide** and **hydrogen**.

$Ca(s) + 2H_2O(l) \rightarrow Ca(OH)_2(aq) + H_2(g)$

Group 2 metals are generally less reactive than the metals in group 1.

The metals in both groups 1 and 2 increase in reactivity down the group.

Barium reacts steadily with cold water to form **barium hydroxide** and **hydrogen**.

$Ba(s) + 2H_2O(l) \rightarrow Ba(OH)_2(aq) + H_2(g)$

There is an increase in reactivity down the group.

Beryllium – **least reactive**
Magnesium
Calcium
Strontium
Barium – **most reactive**

PROGRESS CHECK

1. One alkaline earth metal does not react with cold water. Which metal is this?
2. What is the product of the reaction between strontium, Sr, and oxygen?
3. Write a symbol equation for the reaction of strontium and oxygen.
4. What are the products of the reaction between strontium and cold water?
5. Write a symbol equation for the reaction of strontium and water.
6. One of the alkaline earth metals is normally stored under paraffin oil like alkali metals. Which metal is this? Explain your choice.

1. Beryllium; 2. Strontium oxide; 3. $2Sr(s) + O_2(g) \rightarrow 2SrO(s)$; 4. Strontium hydroxide and hydrogen; 5. $Sr(s) + 2H_2O(l) \rightarrow Sr(OH)_2(aq) + H_2(g)$; 6. Barium. It is an extremely reactive alkaline earth metal.

Explaining the difference in reactivity

Table 7.6 below gives the atomic radii and the electron arrangement of atoms of the elements in group 2 of the periodic table.

Down group 2 the atoms increase in size. This is a similar trend in group 1

Element	Atomic radius in arbitrary units	Electron arrangement
Beryllium	*112*	*2,2*
Magnesium	*160*	*2,8,2*
Calcium	*197*	*2,8,8,2*
Strontium	*215*	*2,8,18,8,2*
Barium	*217*	*2,8,18,18,8,2*

When alkaline earth metals react , each atom loses two electrons to form an ion with a 2+ charge.

e.g. $Mg \rightarrow Mg^{2+} + 2e^-$

Alkaline earth metals (group 2) are less reactive than alkali metals (group 1) because more energy is required to lose two electrons than to lose one electron.

All elements in group 2 have atoms containing two electrons in the outer shell. These two electrons are further from the nucleus. These electrons are lost more easily the further they are away from the nucleus. This is because there is a reduced force of attraction between the nucleus and the outer electrons.

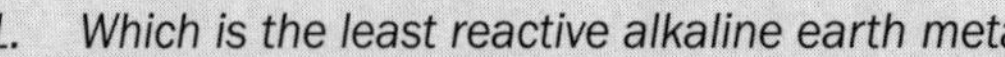

LEARNING SUMMARY

1. *Which is the least reactive alkaline earth metal?*
2. *Write an ionic equation for the change when a strontium atom reacts.*
3. *The table compares some differences between a magnesium atom and a barium atom.*

	Magnesium atom	Barium atom
No. of protons in nucleus	*12*	*56*
Distance of outer electrons from nucleus in arbitrary units	*160*	*217*
Number of electron shells	*3*	*6*

Which one of the differences does not explain the increased reactivity of barium?
Radium is another element in group 2.

4. *How does radium compare in reactivity with the other elements in group 2?*

1. Beryllium; 2. Sr → $Sr^{2+} + 2e^-$; 3. No. of protons in the nucleus; 4. Most reactive.

7.3 Noble gases

LEARNING SUMMARY

After studying this section you should be able to:

- ***understand the trend of properties like boiling point and density for noble gases***
- ***enumerate some of their uses.***

Properties and uses of the noble gases

The noble gases are a family of unreactive gases in group 0 of the periodic table. They were not known when *Mendeleev* devised the first periodic table.

The reason they were not discovered earlier is they are very unreactive. Until about 40 years ago it was believed that they never reacted. We now know that they form some compounds, e.g. xenon tetrafluoride, XeF_4.

Table 7.7 gives some information about noble gases.

Element	Symbol	Boiling point (°C)		Density (g/dm^3)	
Helium	*He*	*−270*	*boiling point increases* ↓	*0.17*	*boiling point increases* ↓
Neon	*Ne*	*−249*		*0.84*	
Argon	*Ar*	*−189*		*1.66*	
Krypton	*Kr*	*−157*		*3.46*	
Xenon	*Xe*	*−112*		*5.46*	
Radon	*Rn*	*−71*		*8.9*	

Table 7.8 gives some uses of noble gases.

Most of these uses rely upon the unreactivity of noble gases.

Noble gas	Use
Helium	*Balloons and airships – less dense than air and not flammable*
Neon	*Filling advertising tubes*
Argon	*Filling electric light bulbs – inert atmosphere for welding*
Krypton and xenon	*Lighthouse and projector bulbs. Lasers*
Radon	*Killing cancerous tumours*

7.4 Transition metals

After studying this section you should be able to:

- ***recall some of the typical properties of transition metals.***
- ***distinguish between transition metal hydroxides on the basis of colour.***

Properties and uses of transition metals

KEY POINT

The transition metals are in a block of metals between groups 2 and 3 in the periodic table.

Iron, nickel and manganese are examples. These metals have a number of features in common including:

Do not confuse the transition metal manganese with magnesium, a metal in group 2.

- **higher melting points**, **boiling points** and **densities** than group 1 metals
- usually **shiny** appearance
- **good conductors of heat and electricity**
- some have **strong magnetic properties**
- often **form more than one positive ion**. For example, iron forms iron(II) ions, Fe^{2+}, and iron(III) ions, Fe^{3+}.
- **compounds are often coloured**. For example, iron(II) sulphate is pale green and iron(III) sulphate is yellow-brown
- transition metals and transition metal compounds are often good **catalysts**. For example, iron is used as the catalyst in the Haber process to produce ammonia.

Transition metal oxides are used to make coloured glazes for pottery.

Transition metals have a wide range of uses, either as pure metals or in mixtures of metals called **alloys**.

Alloys have better properties for most uses than pure metals.

Pure metals are used for electrical conductors as pure metals conduct electricity better.

Steel is an alloy of **iron with a small percentage of carbon**. It is used for making car bodies, ships and bridges. Steel rods are also used to reinforce concrete. **Stainless steel** contains other transition metals such as **nickel** and **chromium**. It is more resistant to corrosion than ordinary steel.

Brass is an alloy of **copper** and **zinc**. It is used for door handles, hinges and decorative ware.

Bronze is an alloy of **copper** and **tin**. It is used to make statues.

Gold and **silver** are used for jewellery, but again they are hardened by alloying with other metals.

Colours of transition metal hydroxides

Transition metal hydroxides are often precipitated when sodium hydroxide is added to a solution of a transition metal compound.

These metal hydroxides have characteristic colours

E.g. $CuSO_4 + 2NaOH \rightarrow Cu(OH)_2 + Na_2SO_4$

Copper(II) sulphate + sodium hydroxide → Copper(II) hydroxide + sodium sulphate

Copper(II) hydroxide is a blue precipitate.

$FeSO_4 + 2NaOH \rightarrow Fe(OH)_2 + Na_2SO_4$

Iron(II) sulphate + sodium hydroxide → Iron(II) hydroxide + sodium sulphate

Iron(II) hydroxide is a dirty green precipitate.

$Fe_2(SO_4)_3 + 6NaOH \rightarrow 2Fe(OH)_3 + 3Na_2SO_4$

Iron(III) sulphate + sodium hydroxide → Iron(III) hydroxide + sodium sulphate

Iron(III) hydroxide is a red-brown precipitate.

PROGRESS CHECK

Here is a list of elements. Use your periodic table to answer these questions.
chlorine helium lithium magnesium titanium

1. *Which element is in period 1?*
2. *Which element is in group 2?*
3. *Which element is an alkali metal?*
4. *Which element is a halogen?*
5. *Which element is a transition metal?*
6. *Which element is a noble gas?*
7. *Which elements have atoms containing two electrons in the outer energy level?*
8. *Which two of these elements are in the same period of the periodic table?*
9. *Sterling silver is an alloy used to make jewellery . Silver is mixed with copper. What are two reasons why sterling silver is better than pure silver for jewellery.*

1. Helium; 2. Magnesium; 3. Lithium; 4. Chlorine; 5. Titanium; 6. Helium; 7. Helium and magnesium; 8. Magnesium and chlorine; 9. Sterling silver is cheaper than pure silver. It is harder than pure silver.

Sample IGCSE questions

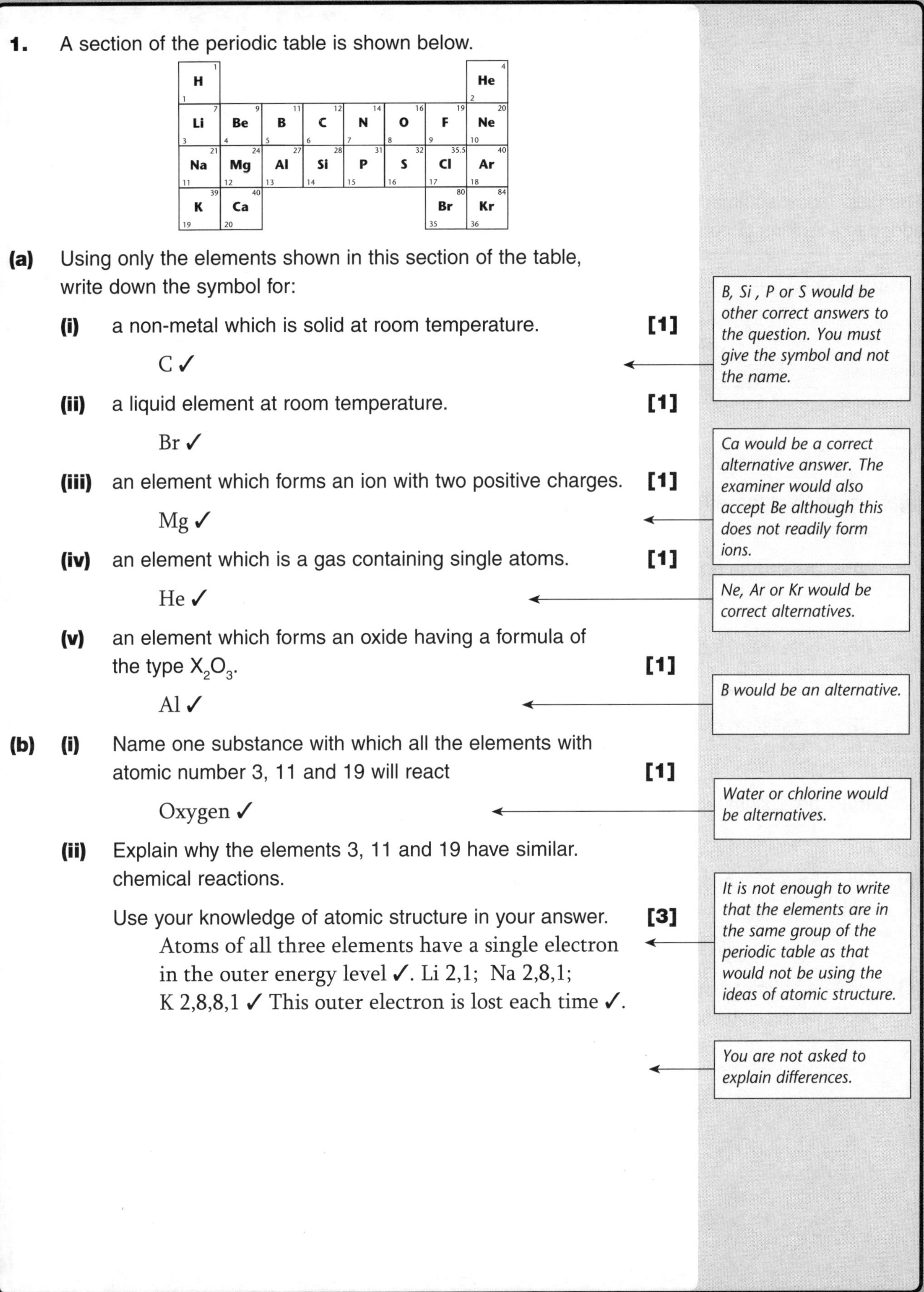

1. A section of the periodic table is shown below.

$^{1}_{1}\text{H}$							$^{4}_{2}\text{He}$
$^{7}_{3}\text{Li}$	$^{9}_{4}\text{Be}$	$^{11}_{5}\text{B}$	$^{12}_{6}\text{C}$	$^{14}_{7}\text{N}$	$^{16}_{8}\text{O}$	$^{19}_{9}\text{F}$	$^{20}_{10}\text{Ne}$
$^{21}_{11}\text{Na}$	$^{24}_{12}\text{Mg}$	$^{27}_{13}\text{Al}$	$^{28}_{14}\text{Si}$	$^{31}_{15}\text{P}$	$^{32}_{16}\text{S}$	$^{35.5}_{17}\text{Cl}$	$^{40}_{18}\text{Ar}$
$^{39}_{19}\text{K}$	$^{40}_{20}\text{Ca}$					$^{80}_{35}\text{Br}$	$^{84}_{36}\text{Kr}$

(a) Using only the elements shown in this section of the table, write down the symbol for:

(i) a non-metal which is solid at room temperature. **[1]**

C ✓

B, Si , P or S would be other correct answers to the question. You must give the symbol and not the name.

(ii) a liquid element at room temperature. **[1]**

Br ✓

(iii) an element which forms an ion with two positive charges. **[1]**

Mg ✓

Ca would be a correct alternative answer. The examiner would also accept Be although this does not readily form ions.

(iv) an element which is a gas containing single atoms. **[1]**

He ✓

Ne, Ar or Kr would be correct alternatives.

(v) an element which forms an oxide having a formula of the type X_2O_3. **[1]**

Al ✓

B would be an alternative.

(b) **(i)** Name one substance with which all the elements with atomic number 3, 11 and 19 will react **[1]**

Oxygen ✓

Water or chlorine would be alternatives.

(ii) Explain why the elements 3, 11 and 19 have similar. chemical reactions.

Use your knowledge of atomic structure in your answer. **[3]**

Atoms of all three elements have a single electron in the outer energy level ✓. Li 2,1; Na 2,8,1; K 2,8,8,1 ✓ This outer electron is lost each time ✓.

It is not enough to write that the elements are in the same group of the periodic table as that would not be using the ideas of atomic structure.

You are not asked to explain differences.

Sample IGCSE questions

2. The order of reactivity of the halogens is:

Fluorine — most reactive
Chlorine
Bromine
Iodine — least reactive

The table below summarises the results of reactions when halogens are added to solutions of potassium halides. Complete the table.

Halogen added	Solutions of		
	Potassium chloride	**Potassium bromide**	**Potassium iodide**
Bromine	✗	✗	✓
Chlorine	✗		
Iodine			✗

(a) For each of these reactions, write **yes** if the reaction takes place or **no** if it does not.

(i) potassium bromide and chlorine **[1]**

Yes ✓

Answering these questions involves using the pattern of reactivity of the halogens given at the start of the question.

(ii) potassium iodide and chlorine **[1]**

Yes ✓

(iii) potassium chloride and iodine **[1]**

No ✓

(iv) potassium bromide and iodine **[1]**

No ✓

(b) What type of reaction is taking place when potassium iodide reacts with bromine? **[1]**

Displacement reaction ✓

(c) Write a balanced symbol equation for the reaction of potassium iodide and bromine. **[3]**

$2KI + Br_2 \rightarrow 2KBr + I_2$ ✓✓✓

The three marks here are for:
1. *the correct formulae of the reactants*
2. *the correct formulae of the products*
3. *balancing correctly.*

Sample IGCSE questions

3. When *Dalton* listed the elements in 1803, he believed that lime was an element.

Calcium was isolated as an element in 1898 by *Sir Humphrey Davy*. He was able to use electricity to extract calcium from calcium chloride.

(a) Suggest why Dalton believed that lime was an element. **[2]**

Lime is very difficult to split up ✓.
Elements are substances that cannot be split up ✓.

This is an Ideas and Evidence question.

(b) Outline how Davy extracted calcium from calcium chloride. **[3]**

Electrolysis ✓ of molten calcium chloride ✓.
Calcium is formed at the cathode (negative electrode) ✓.

(c) Davy added a small piece of calcium to cold water.

(i) Describe his observation. **[4]**

The calcium sank ✓. It fizzed as it reacted with water ✓. A colourless gas was produced ✓. A colourless but slightly cloudy solution was formed ✓.

Calcium sinks because it has a greater density than water i.e. 1 g/dm^3.

(ii) Write a balanced symbol equation for the reaction. **[3]**

$Ca + 2H_2O \rightarrow Ca(OH)_2 + H_2$ ✓✓✓

One mark for correct formulae of reactants, one for correct formulae of products and one mark for correct balancing.

Exam practice questions

1. **(a)** How do group 2 metals compare in reactivity with group 1 metals? Explain this in terms of the structure of the atoms. **[3]**

(b) How does the reactivity of group 2 metals change down the group? Explain this in terms of the structure of the atoms. **[3]**

2. This question is about the elements in group 2 of the periodic table. These elements are called alkaline earth metals.

The reactions of group 2 metals with water.

Magnesium only reacts well with water when heated in steam:

$$Mg(s) + H_2O(g) \rightarrow MgO(s) + H_2(g)$$

Calcium reacts vigorously with cold water:

$$Ca(s) + 2H_2O(l) \rightarrow Ca(OH)_2(aq) + H_2(g)$$

Barium reacts more vigorously than calcium:

$$Ba(s) + 2H_2O(l) \rightarrow Ba(OH)_2(aq) + H_2(g)$$

(a) **(i)** How does the reactivity of alkaline earth metals change down group 2? **[1]**

(ii) Explain this difference in reactivity. Use ideas of atomic structure in your answer. **[4]**

(b) **(i)** Write the formula of calcium chloride. **[1]**

(ii) Describe how a sample of calcium chloride could be produced from calcium hydroxide. **[5]**

Chapter 8

Chemical reactions

The following topics are covered in this section:

- ***Speed of reactions***
- ***Reversible reactions***
- ***Redox***

8.1 Speed of reactions

LEARNING SUMMARY

After studying this section you should be able to:

- ***recall the conditions that can be altered to change the rate of a reaction***
- ***describe an experiment to demonstrate the effect of changing one of the conditions***
- ***explain, using ideas of particles, why changing a condition alters the rate of reaction***
- ***describe and explain the use of enzymes in industrial processes.***

Reactions at different rates

There are chemical reactions that take place very quickly and ones that take place very slowly.

When a lighted splint is placed in a mixture of hydrogen and air, an explosion takes place and a squeaky pop is heard. This reaction is over in a tiny fraction of a second. It is a very fast reaction.

Candidates often confuse rate and time. If a reaction takes longer, the rate decreases.

A limestone building reacts with acidic gases in the air. This reaction takes hundreds of years before the effects can be seen. This is a very slow reaction.

$$\textbf{Rate of reaction} \propto \frac{\textbf{1}}{\textbf{time}}$$

For practical reasons, reactions used in the laboratory for studying rate of reaction must not be too fast or too slow.

Having selected a suitable reaction, it is necessary to find a change that can be observed during the reaction. An estimate of the rate of reaction can be found from the time for a measurable change to take place.

Measuring the volume of gas at intervals

Some of the easiest reactions to study in the laboratory are those where a gas is evolved. The reaction can be followed by measuring the volume of gas evolved over a period of time using the apparatus in **Fig. 8.1**.

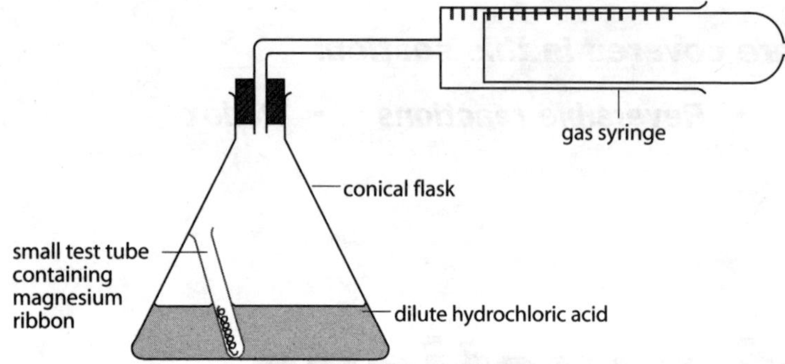

Fig. 8.1 Studying the reaction between magnesium and dilute hydrochloric acid

A good example is the reaction of magnesium with dilute hydrochloric acid:

$$Mg(s) + 2HCl(aq) \rightarrow MgCl_2(aq) + H_2(g)$$

magnesium + hydrochloric acid →
magnesium chloride + hydrogen

It is important to keep the reactants separate whilst setting up the apparatus so that the starting time of the reaction can be measured accurately.

Fig. 8.2 shows a typical graph obtained for the reaction between dilute hydrochloric acid and magnesium.

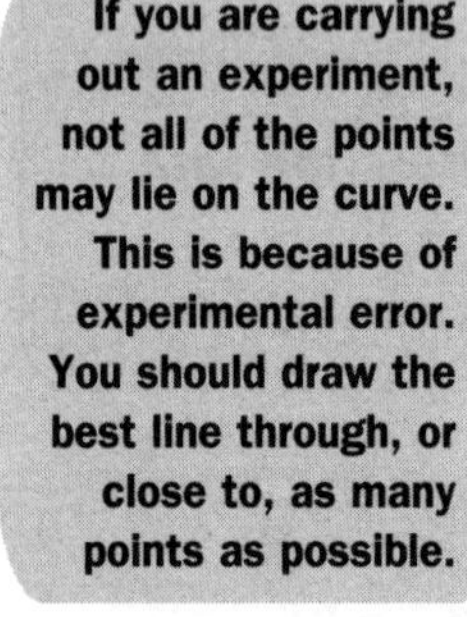

If you are carrying out an experiment, not all of the points may lie on the curve. This is because of experimental error. You should draw the best line through, or close to, as many points as possible.

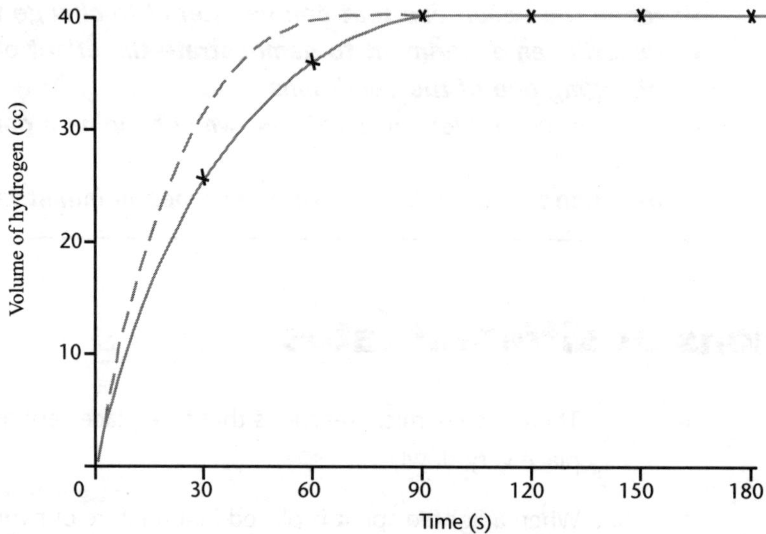

Fig. 8.2 A graph of volume of hydrogen collected at intervals

The dotted line shows the graph for a similar experiment using the same quantities of magnesium and hydrochloric acid, but with changed conditions so that the reaction is slightly faster. The **rate of the reaction is greatest** when the graph is **steepest**, i.e. at the start of the reaction. The reaction is finished when the graph becomes **horizontal**, i.e. there is no further increase in the volume of hydrogen.

It is often possible to follow the course of similar reactions by measuring the loss of mass during the reaction due to escape of gas.

For example, if calcium carbonate and hydrochloric acid are used, the loss of mass of calcium carbonate is significant. However, with magnesium and hydrochloric acid, the loss of mass is very small.

There are ICT opportunities here. The volume of carbon dioxide collected or the mass loss can be found by using a computer.

Other suitable changes that can be measured include:

- **colour changes**
- **formation of a precipitate**
- **time taken for a given mass of solid to react**
- **pH changes**
- **temperature changes.**

Factors affecting rate of reaction

Table 8.1 compares some of the factors that affect the rate of chemical reactions.

Factor	*Reactions affected*	*Change made in conditions*	*Effect on rate of reactions*
Temperature	*All*	*Increase 10 °C* *Decrease 10 °C*	*Approx. doubles rate* *Approx. halves rate*
Concentration	*All*	*Increase in concentration of one of the reactants*	*Increases the rate of reaction*
Pressure	*Reactions involving mixtures of gases*	*Increase the pressure*	*Greatly increases the rate of reaction*
Light	*Wide variety of reactions including reactions with mixtures of gases including chlorine or bromine*	*Reaction in sunlight or uv light*	*Greatly increases the rate of reaction*
Particle size	*Reactions involving solids and liquids, solids and gases, or mixtures of solids*	*Using one or more solids in a powdered form*	*Greatly increases the rate of reaction*
Using a catalyst	*Adding a substance to a reaction mixture*	*A specific substance which speeds up the reaction without being used up*	*Increases the rate of reaction*

Explaining different rates using particle model

Before looking how the rate of reaction can be changed by altering one factor in **Table 8.1**, we must first look at what happens to particles in a reaction.

Particles in solids, liquids and gases are **moving**. This movement is much greater in gases than in liquids and in liquids more than in solids.

In a reaction mixture, the particles of the reactants **collide**. Not every collision leads to reaction. Before a reaction occurs, the particles must have a **sufficient amount of energy**. This is called the **activation energy**. If a collision between particles can produce sufficient energy, i.e. if they collide fast enough and in the right direction, a reaction will take place. Not all collisions will result in a reaction.

A reaction is speeded up if the number of collisions is increased.

Fig. 8.3 shows an energy level diagram of a typical reaction.

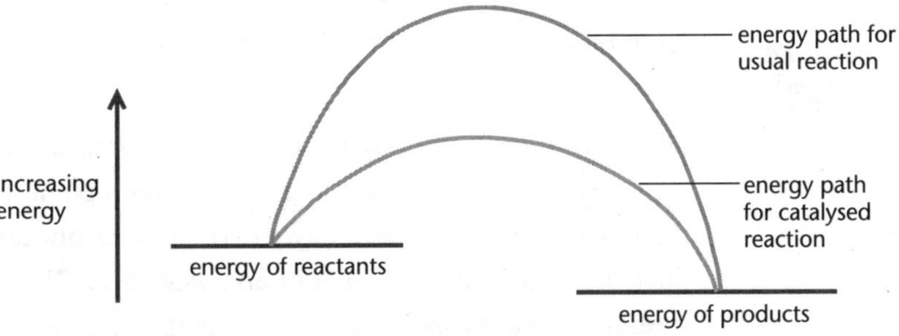

Fig. 8.3

Increasing the concentration

If concentration is increased, there are more collisions between particles and so there are more collisions leading to reaction and the reaction is faster.

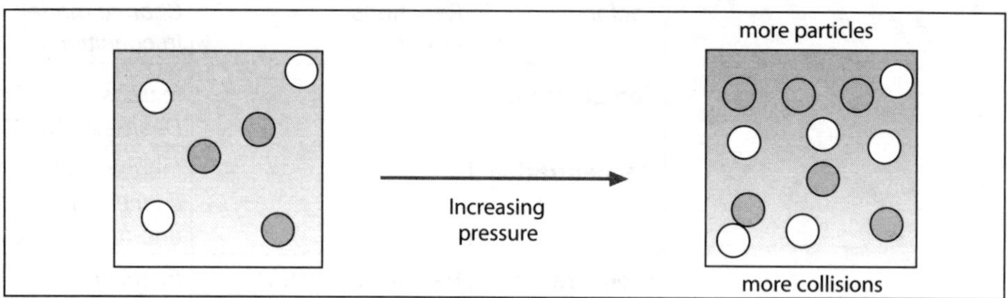

Fig. 8.4 Effect of pressure on reaction rate

Increasing the pressure can be explained in the same way, because increasing the pressure of a mixture of gases increases the concentration by forcing the particles closer together.

Increasing the temperature

Increasing the temperature makes the particles move **faster**. This leads to **more collisions**. Also, the particles have more kinetic energy, so more collisions will lead to reaction. Using sunlight or ultraviolet light has the same effect as increasing temperature.

Using smaller pieces of solid

When one of the reactants is a solid, the reaction must take place on the surface of the solid. By breaking the solid into smaller pieces, the **surface area** is **increased**, giving a greater area for collisions to take place and hence causing an increase in the rate of reaction.

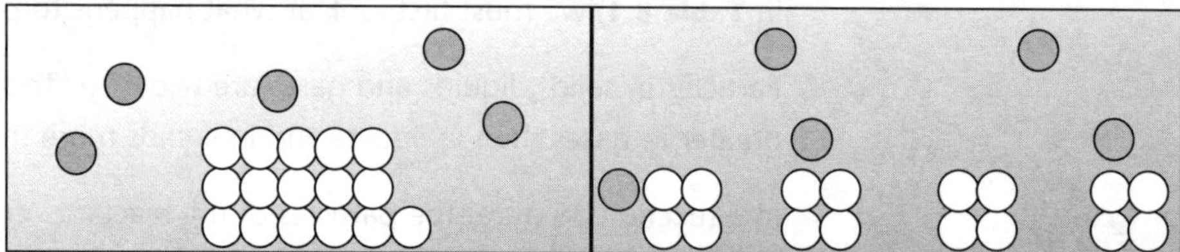

Smaller pieces have a large surface area. More collisions so faster reaction

Fig. 8.5 Effect of surface area on reaction rate

This can cause problems in certain cases such as in flour mills and in coal mines, where the large surface areas of flour and coal dust respectively, react with oxygen present in the air and even a tiny spark can trigger an explosion due to very high rate of reaction.

Effect of light

(The Greek word 'photo' means light).

There are certain chemical reactions which can take place only in the presence of light.The most common examples are **photosynthesis** and **photographic films**.

Photosynthesis is the process in which water absorbed by roots of the plants reacts with carbon dioxide taken in by green plant leaves, in the presence of light, to form glucose and oxygen.The rate of photosynthesis increases with the increase in the intensity of light and vice versa. The green pigment chlorophyll, present in plant leaves, acts as a catalyst in the process.

Carbon dioxide + Water + light energy $\xrightarrow{\text{Chlorophyll}}$ Glucose + Oxygen

In conventional photography, the photographic film used is a transparent plastic strip coated with emulsion containing tiny crystals of silver bromide (AgBr). When exposed to light, the silver cations present in silver bromide crystal accept an electron from the bromide ions (Br^-) and silver atoms are produced.

Silver ion (Ag+) + electron (e-) $\xrightarrow{\text{Light}}$ Silver Atom (Ag)

The amount of silver deposited depends on the intensity of light falling on the photographic film and thus, depending on the light reflected by various parts of the subject being photographed, the amount of silver deposited in different portions of the film is different. This is how the details of the subject are captured on film and that is why the lighter shades appear darker and vice versa in *negatives* of photographs.

Using a catalyst

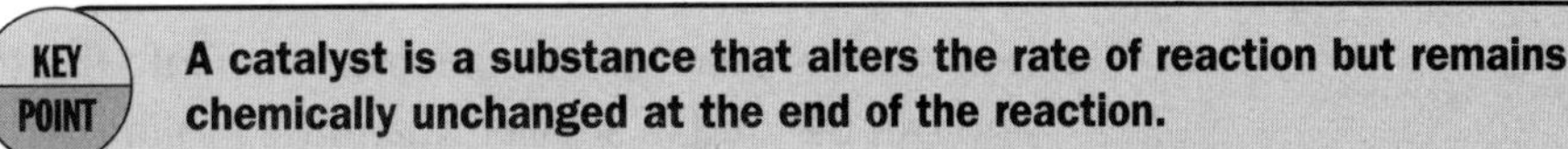

KEY
POINT
A catalyst is a substance that alters the rate of reaction but remains chemically unchanged at the end of the reaction.

Catalysts usually speed up reactions. A catalyst which slows down a reaction is called a negative catalyst or **inhibitor**.

Manganese(IV) oxide catalyses the decomposition of hydrogen peroxide into water and oxygen:

$$2H_2O_2(aq) \rightarrow 2H_2O(l) + O_2(g).$$

Catalysts are often transition metals or transition metal compounds.
The catalyst provides a **surface** where the reaction can take place.

Using a catalyst lowers the activation energy for the reaction. This implies that more collisions have sufficient energy for reactions to take place.

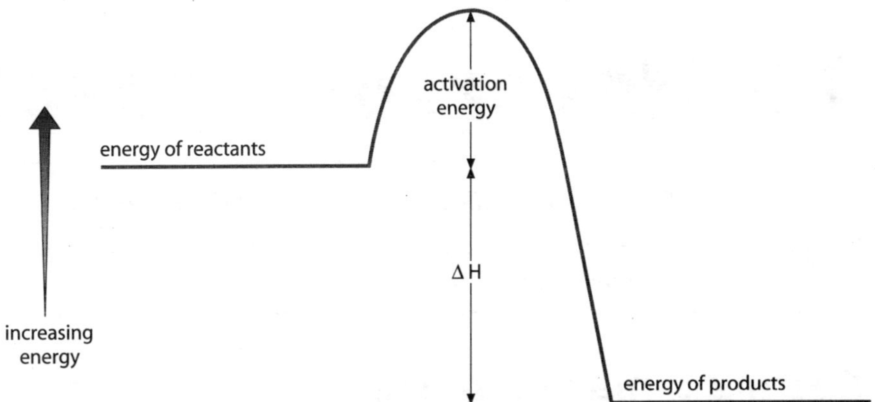

Fig. 8.6 Activation energy

Enzymes

Enzymes are **biological catalysts**. Hydrogen peroxide is decomposed into water and oxygen by an enzyme in fruits and vegetables. Catalase is a protein. Unlike chemical catalysts, such as manganese(IV) oxide, catalase works only under particular conditions. It works best at 37°C. At higher temperatures the protein structure is permanently changed (**denatured**): it no longer decomposes hydrogen peroxide.

Enzymes are used in many industrial processes like:

- fermentation of solutions of starch and sugar using enzymes in yeast to produce beer and wine
- making cheese and yogurt by the action of enzymes on milk
- enzymes (proteases and lipases) in washing powders break down protein stains in cold or warm water
- soft-centred chocolates are made by injecting hard-centred chocolates with the enzyme invertase
- isomerase is used to turn glucose syrup into fructose syrup. This is sweeter and can be used in smaller quantities in slimming products.

Successful enzyme processes

- **stabilise the enzyme, so it works for a long time**
- **trap the catalyst**
- **are continuous.**

Use ideas of rate of reaction to explain each of the following observations.

1. *Mixtures of coal dust and air in coal mines can explode, but lumps of coal are difficult to set alight.*
2. *Milk takes longer to sour when kept in a refrigerator than outside.*
3. *Vegetables cook faster in a pressure cooker.*
4. *Mixtures of methane and chlorine do not react in the dark but react in sunlight.*
5. *Some adhesives are sold in two tubes. The contents of the two tubes have to be mixed before the glue sets.*
6. *Chips fry faster in oil than potatoes cook in boiling water.*

1. Coal dust has a very large surface area. The reaction is speeded up; 2. Lower temperature slows down the rate of souring; 3. In a pressure cooker, increasing the pressure causes an increase in temperature and speeds up the reaction; 4. Light provides initial energy to speed up the reaction; 5. One tube contains a catalyst. Mixing the two tubes speeds up the setting; 6. Oil is at a much higher temperature than boiling water so reactions are faster.

8.2 Reversible reactions

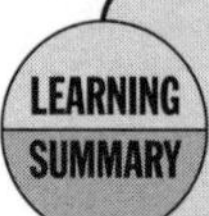

After studying this section you should be able to:

- ***use given data to explain how yields can be maximised in industrial processes involving reversible reactions.***

Getting a maximum yield

In a non-reversible reaction, it is possible to get 100% conversion of the reactants to products.

For example, in the combustion of 24 g of magnesium in excess oxygen, 40 g of magnesium oxide can be formed.

$2Mg + O_2 \rightarrow 2MgO$

In practice, however, complete conversion to products (called 100% yield) is rarely obtained. This is because some reactants and products are lost or other side-reactions take place.

If a reaction is **reversible**, 100% conversion of reactants to products is much more difficult. The amount converted (called **yield**) depends upon conditions.

The reaction of iron with steam is represented by the equation:

$$3Fe(s) + 4H_2O(g) \rightleftharpoons Fe_3O_4(s) + 4H_2(g)$$

If iron and steam in heated is a closed container, **equilibrium** is set up.

This means that the concentrations of the two reactants and the two products remain unchanged, provided conditions are constant.

Apparently, the reaction has stopped. In reality, this is not the case.

The forward and reverse reactions are still taking place, but they are taking place at the same rate.

No more products will be obtained if the mixture is left, and 100% conversion is impossible.

It is possible to get greater conversion of reactants to products, if reactions do not take place in closed containers.

If steam is passed over heated iron in such a way that the hydrogen produced escapes, more steam will react with the iron until all of the iron has reacted.

8.3 Redox

After studying this section you should be able to:

- ***explain oxidation in terms of electron transfer and recall the term redox reaction.***

Redox reactions

Oxidation and reduction reactions occur together. If one substance is oxidised another substance is reduced.
A reaction where oxidation and reduction are taking place is called a redox reaction.

You should know that a reaction where oxygen is gained or hydrogen is lost is called an oxidation reaction. A reaction where oxygen is lost or hydrogen is gained is called a reduction reaction.

Example: if a mixture of lead(II) oxide and carbon are heated together, the following reaction takes place:

$PbO(s) + C(s) \rightarrow Pb(s) + CO(g)$

lead(II) oxide + carbon → lead + carbon monoxide

In this reaction lead(II) oxide is losing oxygen and carbon is gaining oxygen. Lead oxide is **reduced** and carbon is **oxidised**. Reduction and oxidation are taking place and this is called a **redox reaction**.

No reaction would take place if the lead(II) oxide was heated alone. Carbon is the substance which is necessary for the reduction to take place because it removes the oxygen. Carbon is called the **reducing agent**.

KEY POINT **A reducing agent is a substance that reduces some other substances but is itself oxidised.**

Similarly, lead(II) oxide is the **oxidising agent**. It supplies oxygen, which is used to oxidise the carbon.

KEY POINT **An oxidising agent is a substance that oxidises some other substances but is itself reduced.**

Common reducing agents include hydrogen, carbon and carbon monoxide.

Common oxidising agents include oxygen, chlorine, concentrated sulphuric acid and concentrated nitric acid.

Oxidation and reduction in terms of electron transfer

A more advanced definition of oxidation and reduction can be made in terms of loss and gain of electrons.

A simple mnemonic to remember. OILRIG oxidation is loss and reduction is gain (of electrons).

KEY POINT **Oxidation is any process where electrons are lost and reduction is any process where electrons are gained.**

Common examples are:

1. Iron(II) to iron(III) ions

$Fe^{2+} \rightarrow Fe^{3+} + e^-$

Each iron(II) ion loses one electron to become an iron(III) ion. This is **oxidation** (loss of electrons).

2. Chlorine to chloride ions

$Cl_2 + 2e^- \rightarrow 2Cl^-$

Chlorine molecules gain electrons to become chloride ions. This is **reduction** (gain of electrons).

Other halogens, bromine and iodine, act in a similar way.

$Br_2 + 2e^- \rightarrow 2Br^-$

$I_2 + 2e^- \rightarrow 2I^-$

Reaction of chlorine and iron(II) chloride:

If chlorine is bubbled through iron(II) chloride solution, iron(II) ions are oxidised to iron(III) ions and chlorine is reduced to chloride ions.

$2FeCl_2(aq) + Cl_2(g) \rightarrow 2FeCl_3(aq)$

$Fe^{2+} \rightarrow Fe^{3+} + e^-$ (1)

$Cl_2 + 2e^- \rightarrow 2Cl^-$ (2)

These are called half-equations. Two half equations can be added together to form an ionic equation.

Multiply equation 1 by 2 and add the two equations together.

$2Fe^{2+} + Cl_2 \rightarrow 2Fe^{3+} + 2Cl^-$

PROGRESS CHECK

1. *In the reaction of iron(II) ions and chlorine, which substance is the oxidising agent and which is the reducing agent?*

When chlorine is bubbled through potassium iodide solution, a displacement reaction takes place and iodine and chloride ions are produced.

2. *Write an ionic equation for this reaction.*
3. *Which substance is oxidised and which substance is reduced? Explain your answer.*

When a piece of zinc is dipped into copper(II) sulphate solution, a displacement reaction takes place.

4. *Write an equation for the reaction taking place.*
5. *Write an ionic equation for the reaction.*
6. *Write two ionic half equations for the changes taking place.*
7. *Explain what is happening in the reaction in terms of oxidation and reduction.*

1. Iron(II) ions are the reducing agent (they are oxidised) and chlorine is the oxidising agent (it is reduced); 2. $Cl_2 + 2I^- \rightarrow 2Cl^- + I_2$; 3. Chlorine is reduced (gains electrons) and iodide ions are oxidised (lose electrons); 4. $Zn + CuSO_4 \rightarrow ZnSO_4 + Cu$; 5. $Zn + Cu^{2+} \rightarrow Zn^{2+} + Cu$; 6. $Zn \rightarrow Zn^{2+} + 2e^-$, $Cu^{2+} + 2e^- \rightarrow Cu$; 7. Zinc is oxidised (loses electrons), copper(II) ions are reduced (gain electrons).

Redox changes can often be observed as significant colour changes.

E.g.: Potassium manganate(VII) is a powerful oxidising agent and gives an intense purple colour in water due to the MnO_4^- ion. When it oxidises something in acidified solution, it changes to an almost colourless manganese(II) ion, Mn^{2+}.

Potassium iodide is a colourless salt which dissolves in water to form a colourless solution. If it is oxidised, a yellow to orange to brown colour develops, as iodine is formed from the colourless iodide ion.

Sample IGCSE questions

1. An experiment was carried out to investigate the rate of reaction between magnesium and sulphuric acid.

0.07 g of magnesium ribbon reacted with excess dilute sulphuric acid.
The volume of gas produced was recorded every 5 seconds.
The results are shown in the table below.

Time in s	Volume in cm^3	Time in s	Volume in cm^3
0	*9*	*25*	63
5	*18*	*30*	67
10	*34*	*35*	69
15	*47*	*40*	*70*
20	*57*	*45*	*70*

(a) On a piece of graph paper, plot these results with the volume of gas on the y-axis.
Draw a smooth curve through the points. **[3]**

Graph fills over half the grid and labelled axes. ✓
Correct plotting ✓
Curve drawn ✓

(b) When is the reaction fastest? **[1]**

At the start of reaction.

(c) How long does it take for 0.07 g of magnesium to react completely? **[1]**

40 s

(d) At what time was 0.02 g of magnesium left unreacted? **[1]**

When 0.05 g had reacted, 50 cm^3 of gas had been given off; from the graph this is after 18 s. ✓

(e) The experiment was repeated using 0.07 g of magnesium powder instead of magnesium ribbon. How would the graph for this reaction compare with the graph you drew? Explain your answers. **[4]**

The graph is steeper ✓
Reaches the same final level ✓
Powder has a larger surface area so reaction is faster ✓
Same mass of magnesium used. ✓

Sample IGCSE questions

2. The reaction of zinc with sulphuric acid is represented by the symbol equation:

$Zn + H_2SO_4 \rightarrow ZnSO_4 + H_2$

(a) Write an ionic equation for this reaction. **[2]**

$Zn + 2H^+ \rightarrow Zn^{2+} + H_2$ ✓✓

There is one mark for the correct formulae and one mark for balancing.

(b) This ionic equation can be represented by two half ionic equations. Write these two ionic equations. **[4]**

$Zn \rightarrow Zn^{2+} + 2e^-$ ✓✓

$2H^+ + 2e^- \rightarrow H_2$ ✓✓

(c) Explain why this is a redox reaction. **[4]**

This definition of oxidation and reduction in terms of loss and gain of electrons covers more examples than simpler definitions do.

Zinc is oxidised ✓ because it loses electrons ✓. Hydrogen ions are reduced ✓ because they gain electrons ✓.

Exam practice questions

1. Some lumps of zinc (5 g) were put into a flask and 100 cm^3 of hydrochloric acid (100 g/dm^3) added. The temperature was maintained at 20 °C.

How would the rate of formation of hydrogen be affected in each of the following changes made with all the other conditions remaining the same? Explain your answers.

In each experiment, the acid is in excess. **[8]**

New condition	Change, if any	Explanation
Use 5 g of powdered zinc		
Use 40 °C		
Use 100 cm^3 of hydrochloric acid (50 g/dm^3)		
Use 100 cm^3 of ethanoic acid (100 g/dm^3)		

2. Some of the factors that can determine the rate of a reaction are concentration, temperature and light intensity.

A small piece of zinc carbonate was added to an excess of hydrochloric acid. The time taken for the carbonate to react completely was measured.

$$ZnCO_3(s) + 2HCl(aq) \rightarrow ZnCl_2(aq) + CO_2(g) + H_2O(l)$$

The experiment was repeated at the same temperature, using pieces of zinc carbonate of the same size but with acid of different concentrations. In all the experiments, an excess of acid was used.

concentration of acid / mol dm^{-3}	8	4	4	....
number of pieces of carbonate	2	2	4	2
time / s	....	160	....	320

(a) Complete the table (assume the rate is proportional to both the acid concentration and the number of pieces of calcium carbonate). **[3]**

(b) Explain why the reaction rate would increase if the temperature is increased. **[2]**

(c) Explain why the rate of this reaction increases if the piece of carbonate is crushed to a powder. **[1]**

(d) Fine powders mixed with air can explode violently. Name an industrial process where there is a risk of this type of explosion. **[1]**

Exam practice questions

3.(a) Sodium chlorate(I) decomposes to form oxygen and sodium chloride. This is an example of a photochemical reaction. The rate of reaction depends on the intensity of the light.

$$2NaClO(aq) \rightarrow 2NaCl(aq) + O_2(g)$$

(i) Describe how the rate of this reaction could be measured. **[2]**

(ii) How could you show that this reaction is a photochemical reaction? **[1]**

(b) Photosynthesis is another example of a photochemical reaction. Glucose and more complex carbohydrates are made from carbon dioxide and water. Complete the equation.

$$6CO_2 + 6H_2O \rightarrow C_6H_{12}O_6 + \text{..........}$$ **[2]**

Chapter 9 Sulphur

The following topics are covered in this section:

- ***Extraction of sulphur***
- ***Contact process***

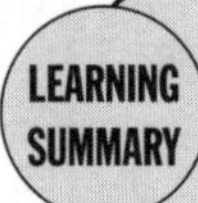

LEARNING SUMMARY

After studying this section you should be able to:

- ***describe the extraction of sulphur and its uses***
- ***understand the process of manufacture of sulphuric acid.***

Extraction of Sulphur

KEY POINT

Sulphur is a non-metallic, yellow coloured solid. It is a very useful element. It is found in crude oil and in metal ores, e.g. copper pyrite, zinc blende, lead sulphide e.t.c. It is also found in elemental state in some places. It is extracted mostly by contact process.

Uses of sulphur

- Sulphur is used to produce sulphuric acid, a very important industrial chemical.
- It is used to produce sulphur dioxide, which is used as a bleach in the manufacture of wood pulp for paper and also as a preservative because of its antibacterial properties.
- Sulphur is used to manufacture several other compounds, such as sulphur dioxide, sulphites and sulphates.
- Being a fungicide and a sterilising agent, it is used as a food preservative and in medicines, e.g. sulpha drugs and skin ointments.
- It is used in the manufacture of matches, fireworks, vulcanised rubber, dyes etc.
- Its salt, sodium thiosulphate (commonly called **hypo**), is used in photography for fixing negatives and prints.

For properties of dilute sulphuric acid as a typical acid refer chapter 11 (Properties – acids)

Contact process

Sulphuric acid is made in industry in a three-stage process.

Stage 1: Burning sulphur or heating sulphide minerals in air:

$$S + O_2 \rightarrow SO_2.$$

Stage 2: Reacting sulphur dioxide with oxygen to produce sulphur trioxide.

$$2SO_2 + O_2 \rightleftharpoons 2SO_3.$$

Stage 3: Absorbing sulphur trioxide to form sulphuric acid:

$$SO_3 + H_2O \rightarrow H_2SO_4.$$

You will not normally be required to recall the details of this process. You may be expected to comment on it when the information is given to you.

Remember, a catalyst does not produce more. It produces the same amount but more quickly.

[In practice, this is done in two stages – the sulphur trioxide is dissolved in concentrated sulphuric acid and then diluted with the required amount of water to make concentrated sulphuric acid.]

Only stage 2 is reversible. Controlling this reaction is the secret for getting the maximum yield in the whole process.

To get the best yield, low temperatures are desirable. However, low temperatures will slow down the reactions. There has to be a **compromise** between getting a good yield and getting the yield quickly. Vanadium(V) oxide can be used as a catalyst to speed up the reaction.

PROGRESS CHECK

1. *What is the name of the industrial process used to manufacture ammonia?*
2. *What is the catalyst in the process producing ammonia?*
3. *What is the name of the industrial process used to make sulphuric acid?*
4. *What are the raw materials in the industrial process producing sulphuric acid?*
5. *Which two gases react together in the catalyst chamber to produce sulphur trioxide?*
6. *What is the catalyst in the process producing sulphuric acid?*
7. *What is the significance of conditions in the catalyst chamber?*

1. Haber process; 2. Iron; 3. Contact process; 4. Sulphur, air and water; 5. Sulphur dioxide and oxygen; 6. Vanadium(V) oxide; 7. The conditions in the catalyst chamber determine the total yield in the process.

Sample IGCSE questions

1. Give two physical properties of sulphur. Where is it found? **[2 + 2]**

Non-metallic ✓, yellow coloured solid ✓. It is found in crude oil ✓ and in metal ores ✓.

2. Give five uses of sulphur. **[5]**

Sulphur is used to produce sulphuric acid, to produce sulphur dioxide (which is used as a bleach in the manufacture of woodpulp for paper and also as a preservative because of its antibacterial properties), to manufacture several other compounds such as sulphites and sulphates, as a fungicide and a sterilising agent. It is also used as a food preservative and in medicines, e.g. sulpha drugs and skin ointments, used in the manufacture of matches, fireworks, vulcanised rubber, dyes etc.

You may include any five of the given uses in your answer.

Exam practice questions

1. In which metal ores is sulphur found? **[2]**

2. Which method is used to remove sulphur impurities from sulphur ores? **[1]**

3. Name the process of extraction of sulphur. **[1]**

4. Which compound of sulphur is used in photography? **[1]**

5. Which compound of sulphur is used as a bleach in the manufacture of woodpulp for paper and also as a preservative? **[1]**

Chapter

10 Chemical changes

The following topics are covered in this section:

- *Energetics of a reaction*
- *Production of energy*

10.1 Energetics of a reaction

LEARNING SUMMARY

After studying this section you should be able to:

- ***recall that during exothermic reactions energy is lost to the surroundings while in endothermic reactions, the energy is taken in from the surroundings***
- ***recall that energy is required to break chemical bonds and energy is released when bonds are formed***
- ***draw energy level diagrams for exothermic and endothermic reactions***
- ***use bond energy data to calculate energy changes in reactions.***

Endothermic and exothermic reactions

There are many examples where energy is either **released** or **taken** in during a chemical reaction.

Exothermic reactions

The burning of carbon in oxygen releases energy. Such a reaction is called an **exothermic reaction**.

The **quantity of energy** released depends upon the **mass** of carbon burned:

$C + O_2 \rightarrow CO_2$

carbon + oxygen → carbon dioxide ($\Delta H = -393.5$ kJ)

This information tells a chemist that burning 12 g of carbon in oxygen produces 393.5 kJ.

ΔH is called the **enthalpy** (or heat) **of reaction**. A **negative** value is used to show energy that is released.

The process is summarised in **Fig. 10.1**.

If two solutions are mixed at room temperature and the temperature rises, an exothermic reaction has taken place.

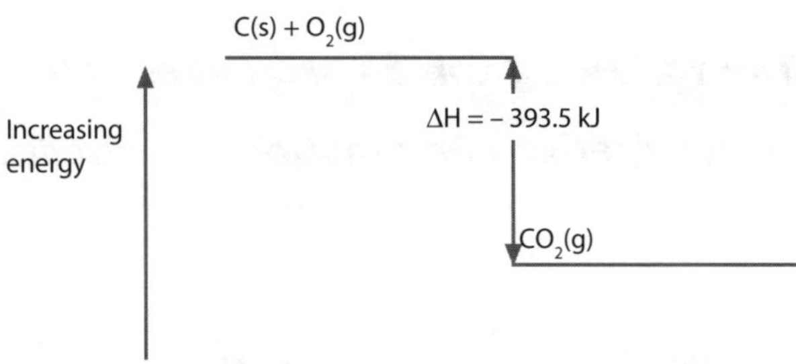

Fig. 10.1 Energy diagram for the complete combustion of carbon

Endothermic reactions

There are some reactions where energy is absorbed from the surroundings during the reaction and the temperature falls. These are called **endothermic** reactions.

For example, the formation of hydrogen iodide from hydrogen and iodine absorbs energy from the surroundings:

$H_2(g) + I_2(g) \rightleftharpoons 2HI(g)$

hydrogen + iodine $\rightleftharpoons$ hydrogen iodide ($\Delta H = +52$ kJ)

ΔH is **positive**, in this case, because the reaction is endothermic. This is summarised in **Fig. 10.2**.

Most reactions are exothermic. At one time scientists thought that only exothermic reactions could take place because they could not find any endothermic ones.

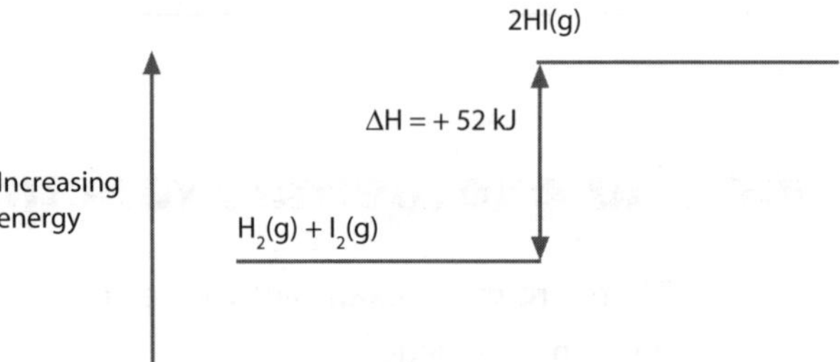

Fig. 10.2 Energy diagram for the formation of hydrogen iodide

Bond making and bond breaking

During a chemical reaction, changes in bonding take place. Whether a particular reaction is exothermic or endothermic depends upon the energy required to break bonds and the energy released during bond formation.

If a reaction is an exothermic reaction:
energy released when bonds form > energy required to break bonds.
The surplus energy raises the temperature of the surroundings.
If a reaction is an endothermic reaction:
energy released when bonds form < energy required to break bonds.
The extra energy needed is taken from the surroundings so the temperature of the surroundings falls.

E.g.

Hydrogen and chlorine react together to form hydrogen chloride:

$H_2(g) + Cl_2(g) \rightarrow 2HCl(g)$.

Using relative atomic masses, 2 g of hydrogen react with 71 g of chlorine to form 73 g of hydrogen chloride.

This can be represented by:

H–H Cl–Cl → H–Cl H–Cl.

We can find out the energy required to break 2 g of hydrogen molecules and 71 g of chlorine molecules by looking up data in a data book. Also, we can find the energy released when 73 g of hydrogen chloride is formed:

Energy required to break H–H bonds in 2 g of hydrogen molecules = +436 kJ.

Energy required to break Cl–Cl bonds in 71 g of hydrogen molecules = +242 kJ.

From a data book, we can find out the energy released when 36.5 g of hydrogen chloride is formed:

When bonds are formed, the value is positive. When bonds are broken the value is negative.

Energy released when forming H–Cl bonds in 36.5g of hydrogen molecules = –431 kJ.

The energy released when forming H–Cl bonds in 71 g (i.e. 2 × 36.5 g) of hydrogen molecules = 2 x (+431) = –862 kJ.

Energy change = (+436) + (+242) – 862 = –184 kJ.

The negative value tells us that the reaction is exothermic.

10.2 Production of energy

LEARNING SUMMARY

After studying this section you should be able to:

- ***enumerate different sources used to produce energy***
- ***understand the meaning of terms like fission and chain reaction***
- ***understand that working of cells involves a redox reaction.***

Burning of fuels - A fuel is a substance which can produce energy, when it undergoes **combustion**. Examples of fuel are wood, coal, gas, oil etc.

Fuels burn in the presence of oxygen to produce heat energy and carbon dioxide (a harmful gas) and water are formed in this process. This can be illustrated as follows.

Fuel + Oxygen ⟶ Carbon dioxide + Water + Heat energy

Hydrogen as fuel

Hydrogen burns in air to produce energy and water is formed in this process. Thus, no harmful gases are generated in the process.

Hydrogen + Oxygen ⟶ Water

$$2H_2 + O_2 \longrightarrow 2H_2O$$

At present, the use of hydrogen as a fuel is limited but it is likely to grow in the future.

Radioactive isotopes as energy sources

Radioactive isotopes such as Uranium-235 (^{235}U) are considered the best possible sources to produce energy in large amounts.

Energy is produced from ^{235}U by a process called **nuclear fission**. This occurs when the unstable nucleus of a radioactive isotope splits up to form smaller atoms and in this process, some of the mass of the isotope is converted into energy.

The fission process is triggered by hitting an atom of Uranium-235 with a neutron. This results in splitting of the uranium atom and production of three more neutrons. The three neutrons split three more atoms which produce nine neutrons and so on, thus initiating a **chain reaction** and producing tremendous amount of energy.

To control the production of this energy, the fission process is conducted inside a nuclear reactor, which has facilities to control the fission process. This control is exercised by pushing boron rods into the reactor, which absorb some of the neutrons, thus slowing down the chain reaction.

The energy thus generated is used to produce steam, which in turn is used to power turbines for generating electricity.

Cells and batteries as energy sources

A cell is a unit which produces electrical energy. The most commonly used cells are **chemical cells**. These cells produce energy by transfer of electrons, which takes place during a chemical reaction via a redox process. (See section 13.1 for detailed information.)

The main advantage of cells and batteries is that they can be constructed in very small sizes and are, therefore, widely used to power portable electronic and electrical devices like torch, radio, watches etc.

1. *When sodium carbonate and calcium chloride solutions are mixed, the temperature of the solution dropped by 4 °C. Is the reaction exothermic or endothermic?*
2. *Are combustion reactions exothermic or endothermic?*
3. *Is energy needed or given out when bonds are broken?*
4. *Is energy needed or given out when bonds are formed?*
5. *Are ΔH values for an exothermic reaction positive or negative?*
6. *Are ΔH values for an endothermic reaction positive or negative?*
7. *Suggest why the following reaction will be endothermic:* $N_2 + O_2 \rightarrow 2NO$.
8. *The breaking of a radioactive nucleus into two or more lighter fractions is called ....*
9. *What is a self-sustaining nuclear reaction called?*
10. *What is a chemical cell?*

1. Endothermic; 2. Exothermic; 3. Needed; 4. Given out; 5. Negative; 6. Positive;
7. Strong N = N and O = O bonds have to be broken and only two NO bonds formed;
8. Nuclear fission; 9. Chain Reaction; 10. System for converting chemical to electrical energy

Sample IGCSE questions

1. What is a fuel? **[2]**

A fuel is a substance which can produce energy when it undergoes combustion. ✓ Examples of fuel are wood, coal, gas, oil etc. ✓

2. How can hydrogen be used as a fuel? **[1]**

Hydrogen burns in air to produce energy ✓ and water is formed, in this process.

3. (a) What is a chain reaction? **[2]**

The fission process is triggered by hitting an atom of uranium-235 with a neutron. This results in splitting of the uranium atom and production of three more neutrons. ✓ The three neutrons further split three more atoms, which produce nine neutrons and so on, thus initiating a chain reaction ✓ and producing limitless energy.

(b) How can we control it? **[1]**

We can control it using boron rods which absorb neutrons. ✓

4. Which process is responsible for generation of energy in **[2]**

(a) a nuclear reactor

Fission reaction ✓

(b) a cell?

Chemical (redox) reaction ✓

Exam practice questions

1. **(a)** What is the difference between endothermic and exothermic reactions? **[2]**

(b) If the energy required to break bonds of reactants is more than the energy released when bonds are formed, then what kind of reaction will it be? **[1]**

2. What is the main advantage of using cells and batteries as source of energy, as compared to other sources? **[2]**

Chapter 11 Acids, bases and salts

The following topics are covered in this section:

- ***Properties***
- ***Preparation of salts***
- ***Identification of ions and gases***

You will know that there are substances called indicators that help in identifying acids and alkalis by undergoing a color change. Litmus, for example, turns red in acids and blue in alkalis.
Acids and alkalis react in neutralisation reactions. This unit studies the topic at greater depth. It includes tests for common positive and negative ions.

11.1 Properties

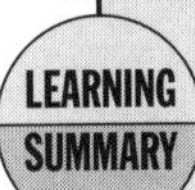

After studying this section you should be able to:

- ***recall the properties of acids and bases***
- ***explain the difference between strong and weak acids/bases***
- ***explain the process of neutralisation and give common examples***
- ***understand the concept of pH scale***
- ***classify oxides as acidic, basic or amphoteric.***

Acids

Acids:

- **are compounds containing hydrogen that can be replaced by a metal**
- **dissolve in water to form hydrogen, H^+ ions**
- **are proton donors.**

An acid with one replaceable hydrogen atom is called a monobasic acid.

The hydrogen in hydrochloric acid, HCl, can be replaced by sodium to form sodium chloride, NaCl.

Table 11.1 below gives a list of names and formulae of some common acids.

acid	formula	
Hydrochloric acid	HCl	*Mineral acids*
Sulphuric acid	H_2SO_4	
Nitric acid	HNO_3	
Ethanoic acid (Acetic acid)	CH_3COOH	*Contained in vinegar*
Ethanedioic acid	$C_2O_4H_2$	*Contained in rhubarb leaves*
Citric acid	$C_3H_8O_7$	*Contained in lemon juice*

Ethanoic acid contains four hydrogen atoms per molecule but is a monobasic acid.

Only one hydrogen atom in ethanoic acid can be replaced, forming sodium ethanoate CH_3COONa.

Properties of acids

Although there are a large number of different acids, there are a number of general chemical reactions common to all acids.

Indicators

Acids turn indicators to their characteristic colours, e.g. litmus turns red.

Fairly reactive metals

KEY POINT **Acids react with fairly reactive metals (e.g. magnesium and zinc) to form a salt and hydrogen gas.**

The equations show state symbols. These are optional unless they are specifically asked for.

e.g. $Mg(s) + H_2SO_4(aq) \rightarrow MgSO_4(aq) + H_2(g)$

magnesium + sulphuric acid → magnesium sulphate + hydrogen

$Zn(s) + 2HCl(aq) \rightarrow ZnCl_2(aq) + H_2(g)$

Zinc + hydrochloric acid → zinc chloride + hydrogen

Nitric acid is an exception. This acid tends to release oxides of nitrogen when it reacts with metals.

Metal oxides

KEY POINT **Acids react with metal oxides to form a salt and water only.**

Metal oxides are called bases.

A blue solution is formed with Copper Sulphate.

In most cases warming is necessary.

e.g. $CuO(s) + H_2SO_4(aq) \rightarrow CuSO_4(aq) + H_2O(l)$

copper(II) oxide + sulphuric acid → copper(II) sulphate + water

$ZnO(s) + 2HCl(aq) \rightarrow ZnCl_2(aq) + H_2O(l)$

Zinc oxide + hydrochloric acid → zinc chloride + water

Metal carbonates

KEY POINT **Acids react with carbonates (or hydrogencarbonates) to form carbon dioxide, a salt and water.**

e.g. $CaCO_3(s) + 2HCl(aq) \rightarrow CaCl_2(aq) + H_2O(l) + CO_2(g)$

calcium carbonate + hydrochloric acid → calcium chloride + water + carbon dioxide

$2NaHCO_3(s) + H_2SO_4(aq) \rightarrow Na_2SO_4(aq) + 2H_2O(l) + 2CO_2(g)$

sodium hydrogencarbonate + sulphuric acid → sodium sulphate + water + carbon dioxide

Metal hydroxides

Metal hydroxides that dissolve in water are called alkalis.

KEY POINT **Alkalis react with acids to form a salt and water only.**

e.g. $NaOH(aq) + HCl(aq) \rightarrow NaCl(aq) + H_2O(l)$

sodium hydroxide + hydrochloric acid → sodium chloride + water

These tests are used to identify the presence of an acid.

These reactions of acids can be represented by **ionic equations**.

e.g. $Mg(s) + 2H^+(aq) \rightarrow Mg^{2+}(aq) + H_2(g)$

$$O^{2-}(s) + 2H^{+}(aq) \rightarrow H_2O(l)$$

$$CO_3^{2-}(s) + 2H^{+}(aq) \rightarrow H_2O(l) + CO_2(g)$$

$$H^{+}(aq) + OH^{-}(aq) \rightarrow H_2O(l)$$

Dry acids do not show acidic properties. Water must be present.

The idea of dissociation of acids came from Arrhenius. These ideas were developed later by Lowry and Bronsted.

Arrhenius' theory did not involve the solvent and so could not explain some examples.

A strong acid is one that is completely ionised. It has got nothing to do with the corrosive action of the acid. Nor is it a measure of the concentration of an acid.

Strong and weak acids

Some acids completely **ionise** when they dissolve in water. These are called **strong acids**.

A solution of a strong acid will have a high concentration of hydrogen ions, H^+.

e.g. sulphuric acid

$$H_2SO_4\,(l) \xrightarrow{\text{water}} 2H^{+}\,(aq) + SO_4^{2-}\,(aq)$$

Other acids do not ionise completely on dissolving in water. Some of the molecules remain **un-ionised** in the solution. These are called **weak acids**.

e.g. $$CH_3COOH\,(l) \overset{\text{water}}{\rightleftharpoons} CH_3COO^{-}(aq) + H^{+}\,(aq)$$

In a solution of ethanoic acid (1 mol/dm^3), there are about four, in every thousand molecules, which get ionised.

Bases

Bases:
- **are chemical opposites of acids**
- **dissolve in water to form hydroxide, OH^- ions**
- **are proton acceptors.**

Bases are also called alkalis.

Table 11.2 below gives a list of names and formulae of some common bases.

base	formula
Sodium hydroxide	*NaOH*
Potassium hydroxide	*KOH*
Ammonia	NH_3
Calcium carbonate	$CaCO_3$

Properties of bases

There are a number of general chemical reactions common to all bases.

Indicators

Bases turn indicators to their characteristic colours e.g. red litmus turns blue.

Acids

KEY POINT **Bases react with acids to form salt and water.**

$$KOH + HCl \longrightarrow KCl + H_2O$$

Gases being dry do not show acidic properties.

In gaseous form, the reaction is different.

$$NH_3\ (g) + HCl\ (g) \longrightarrow NH_4Cl\ (s)$$

Strong and weak bases

Bases which ionise completely, when dissolved in water, are called **strong bases**.

A solution of a strong base will have a high concentration of hydroxide ions, OH^-.

The number of OH^- ions produced in this reaction is very high.

e.g. Sodium hydroxide

$$NaOH\ (s) + H_2O\ (l) \xrightleftharpoons{water} KCl\ (s) + H_2O(l)$$

Bases which ionise only partially when dissolved in water are called **weak bases**.

The number of OH^- ions produced is very low.

e.g. Ammonia

$$NH_3\ (g) + H_2O\ (l) \rightleftharpoons NH_4^+\ (aq) + OH^-\ (aq)$$

PROGRESS CHECK

1. *Which of the acids in the list is not a strong acid?*
   ethanoic acid hydrochloric acid nitric acid sulphuric acid
2. *Which acid reacts to form sulphates?*
3. *Which acid reacts to form nitrates?*
4. *Which acid reacts to form chlorides?*

Carbonic acid is a weak dibasic acid, H_2CO_3.

5. *Write an equation showing the ionisation of carbonic acid.*
6. *What is meant by the term weak acid?*
7. *What is a dibasic acid?*
8. *What salts are produced by carbonic acid?*

Finish the following word equations.

9. *dilute acid + fairly reactive metal →* ____________ + ____________
10. *dilute acid + metal oxide →* ____________ + ____________
11. *dilute acid + metal carbonate →* ____________ + ____________ + ____________
12. *dilute acid + an alkali →* ____________ + ____________

1. Ethanoic acid; 2. Sulphuric acid; 3. Nitric acid; 4. Hydrochloric acid;
5. $H_2CO_3 \rightleftharpoons 2H^+ + CO_3^{2-}$; 6. Only partially ionised; 7. Two replaceable hydrogen atoms;
8. Carbonates; 9. salt + hydrogen; 10. salt + water; 11. salt + water + carbon dioxide;
12. salt + water.

Neutralisation

KEY POINT **Neutralisation is the reaction of acid and alkali, in the correct proportions, to produce a neutral substance.**

Examples of neutralisation

1. Soil that is too acidic is not as fit to grow crops as neutral soil.
   Slaked lime (calcium hydroxide) or **limestone** (calcium carbonate) can be added to the soil to neutralise it.

A neutral solution contains equal concentrations of H^+ and OH^- ions. An acidic solution contains an excess of H^+ ions while an alkaline solution contains an excess of OH^- ions.

2. **Hydrochloric acid** in the stomach helps in the digestion of food. **Indigestion** is caused by excess acid. The pain can be relieved by taking a **weak alkali** such as sodium hydrogencarbonate or magnesium hydroxide.
3. Insect bites and stings involve an injection of a small amount of chemical into the skin. These chemicals cause irritation. Nettle stings, bee stings and ant bites involve methanoic acid being injected into the skin. Wasp stings involve injection of an alkali into the skin. The irritation can be removed by neutralisation of the acid or alkali.
4. Coal-fired power stations produce sulphur dioxide, which can produce **acid rain**. The sulphur dioxide can be removed from the waste gases before they escape into the atmosphere. Limestone removes sulphur dioxide from the waste gases.
5. Acid rain, caused by sulphur dioxide escaping into the atmosphere, can make lakes and rivers acidic. This can affect organisms in the water such as killing fish. Blocks of limestone, put into the water, can reduce the acidity.

KEY POINT

Neutralisation can be summarised by the ionic equation:
$H^+(aq) + OH^-(aq) \rightarrow H_2O(l)$

Relative acidity and alkalinity

Concentration is different from strength. It indicates the relative proportion of water and acid/base in an aqueous solution.

The degree to which acids and bases ionise in solution determines the relative **strength** of acids and bases. To measure the acidity/alkalinity of a substance, a scale called pH scale has been developed.The pH scale runs from 0 to 14.

KEY POINT

A substance with pH of 7 is called neutral e.g. water.

The pH scale was developed by Scandinavian chemist *Soren Sorenson.*

If the value of pH is **less than 7**, the solution is acidic. Lesser the pH value, more acidic is the solution i.e. more is its relative acidic strength.

e.g. Hydrochloric acid is a stronger acid compared to lemon juice – so its pH value is lower.

If the value of pH is **more than 7**, the solution is basic. More the pH value, more basic is the solution i.e. more is its relative basic strength.

e.g. Sodium hydroxide solution is a stronger base compared to lime water–so its pH value is higher.

Litmus, one of the oldest indicators, turns red in acid and blue when dipped in alkali.

To check whether a substance is acidic or alkaline, substances called **indicators** are used. Indicators change colour on coming in contact with acids or alkalis.

To obtain an idea of how acidic or alkaline a substance is, an indicator called **universal indicator,** is used.

Other indicators include methyl red, phenolphthalein etc., each being useful for a particular range of acidity.

Universal indicator is a mixture of several indicators and its colour changes with change in acidity or alkalinity of the solution, in which it is kept. The colour shown by this indicator can be matched against the pH scale to determine the relative degree of acidity/alkalinity of a substance.

All the strips between 1 to 14 have different colour shades.

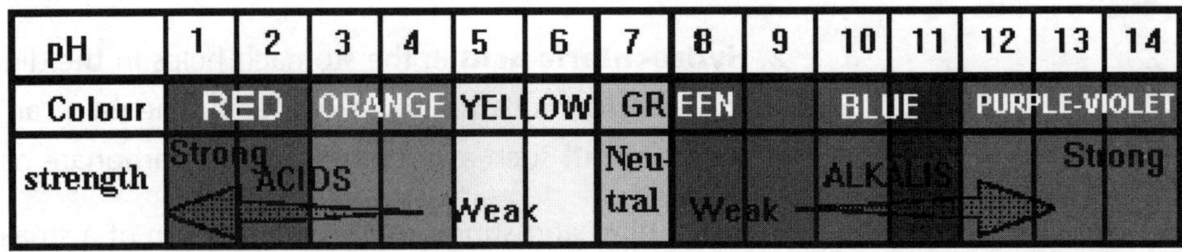

pH	1	2	3	4	5	6	7	8	9	10	11	12	13	14
Colour	RED		ORANGE		YELLOW		GREEN			BLUE		PURPLE-VIOLET		
strength	Strong	ACIDS			Weak		Neutral	Weak		ALKALIS			Strong	

Fig 11.1 pH scale

Definition of acids and bases – proton transfer

Historically, many attempts have been made by eminent chemists viz. *Antoine Lavoisier, Humphrey Davy, Justus van Liebig* and *Svante Arrhenius,* to define acids and bases.

The most general theory of acids and bases was proposed independently by the chemists *Johannes Bronsted* and *Thomas Lowry*.

According to Bronsted–Lowry theory:

- **an acid is a H^+ ion (or Proton) donor**
- **a base is a H^+ ion (or Proton) acceptor.**

In an aqueous solution of an acid, the acid acts as an H^+ ion donor and water behaves as a base, accepting H^+ ions.

E.g. $$\underset{\text{(Acid)}}{HCl\,(aq)} + \underset{\text{(Base)}}{H_2O\,(l)} \longrightarrow H_3O + (aq) + Cl$$

A substance acting both as an acid and a base is called ***amphoteric*** e.g. Water.

Similarly in an aqueous solution of a base, water acts as an H^+ ion donor(an acid) and ammonia behaves as a base, accepting H^+ ions.

E.g. $$\underset{\text{(Base)}}{NH_3\,(aq)} + \underset{\text{(Acid)}}{H_2O\,(l)} \longrightarrow NH_4^+\,(aq) + OH^-\,(aq)$$

The quantity of base needed to neutralise a weak acid is the same as that required for a strong acid of the same concentration.

Some soils are naturally acidic but may become more acidic due to acid rain or contamination by acidic gases,liquids or solids. Increase in acidity of soil makes it unsuitable for plant growth. To overcome this problem, acidic soil has to be neutralised by a suitable base.

The most commonly used base for neutralising acidic soil is powdered limestone ($CaCO_3$). The reaction is as follows:

$$\underset{\text{(From }CaCO_3)}{CO_3^{2-}(s)} \;+\; \underset{\text{(From acidic soils)}}{2H^+\,(aq)} \longrightarrow CO_2(s) + H_2O(l)$$

Classification of Oxides

Oxides are classified as acidic, basic, neutral and amphoteric.

The metallic oxides, e.g. copper oxide (CuO), zinc oxide (ZnO) etc. are mostly basic in nature and they react with acids to form a salt and water.

The non-metallic oxides e.g. SO_2 , NO_2 etc. are mostly acidic in nature and they react with bases to form a salt and water.

Some oxides, such as water(H_2O), are neutral i.e. they are neither acidic or basic.

Some other oxides, such as aluminium oxide (Al_2O_3) and zinc oxide(ZnO), are *amphoteric* i.e. they have both acidic and basic properties.

E.g.

$$\underset{\text{(base)}}{ZnO\ (s)} + \underset{\text{(acid)}}{2HCl\ (aq)} \longrightarrow ZnCl_2\ (aq) + H_2O$$

$$\underset{\text{(acid)}}{ZnO\ (s)} + \underset{\text{(base)}}{2NaOH\ (aq)} + \underset{\text{(Neutral)}}{H_2O} \longrightarrow Na_2Zn(OH)$$

In one reaction, ZnO acts as a base and in the other, as an acid.

PROGRESS CHECK

1. *Why is sodium hydrogencarbonate used to treat bee stings but vinegar used to treat wasp stings?*
2. *Write an equation for the neutralisation of hydrochloric acid in the stomach by sodium hydrogencarbonate.*
3. *Write an equation for the neutralisation of hydrochloric acid in the stomach by magnesium hydroxide.*
4. *Sodium hydroxide, NaOH, is a* **strong** *alkali and ammonium hydroxide, NH_4OH, is a* **weak** *alkali. Explain the meaning of the terms stong and weak.*

1. Bee stings are acidic and need an alkali to neutralise them. Wasp stings are alkaline and need an acid; 2. $NaHCO_3(s) + HCl(aq) \rightarrow NaCl(aq) + H_2O(l) + CO_2(g)$; 3. $Mg(OH)_2(s) + 2HCl(aq) \rightarrow MgCl_2(aq) + 2H_2O(l)$; 4. Sodium hydroxide is completely ionised, ammonium hydroxide is only partly ionised.

11.2 Preparation of salts

Preparing soluble salts

LEARNING SUMMARY

After studying this section you should be able to:

- ***describe how soluble and insoluble salts can be prepared.***

Soluble salts are salts that readily dissolve in water.

Common soluble salts are:

- all metal nitrates
- all metal chlorides except silver chloride and lead chloride
- all metal sulphates except lead sulphate and barium sulphate. (Calcium sulphate is only sparingly soluble)
- sodium, potassium and ammonium carbonates.

Method used to produce a particular salt depends upon various factors – availability, cost and the speed of the reaction (not too fast or not too slow).

KEY POINT

Four methods of preparing soluble salts are:

- **Acid + metal**
- **Acid + metal oxide**
- **Acid + metal hydroxide**
- **Acid + metal carbonate.**

The method used in each case is the same and is summarised in **Fig. 11.2.**

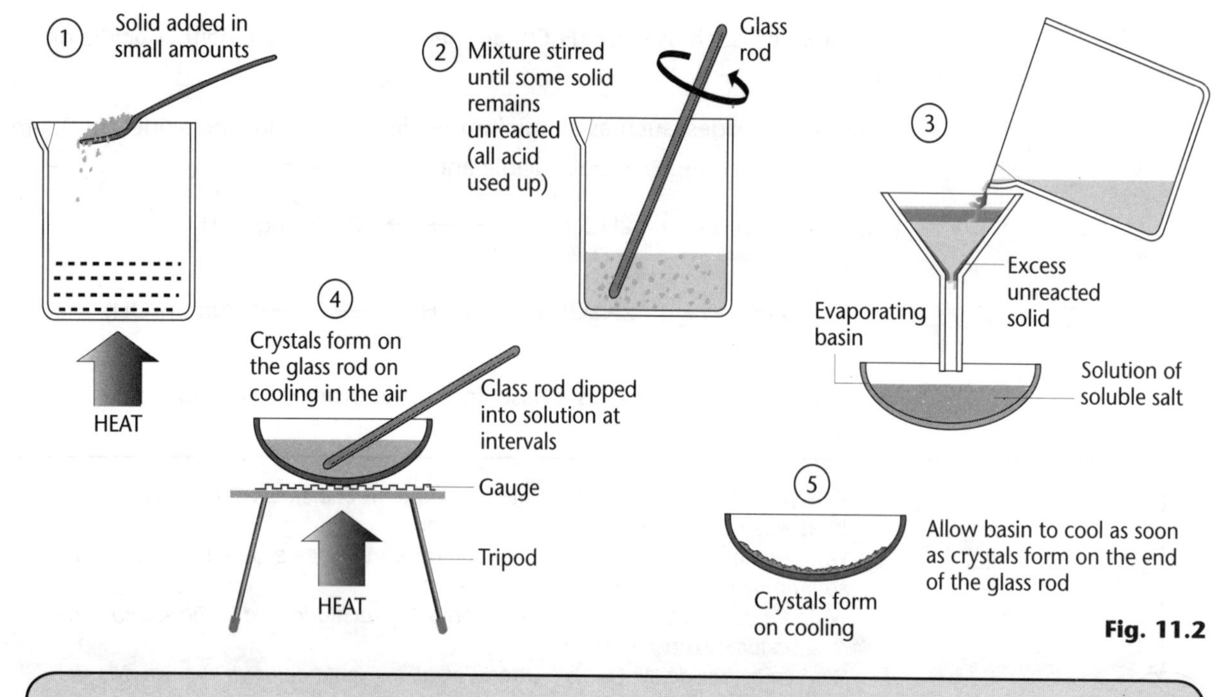

Fig. 11.2

PROGRESS CHECK

1. *Which salts in the list are soluble in water?*
   sodium sulphate   sodium carbonate   lead carbonate   silver nitrate   barium chloride
2. *Magnesium sulphate can be prepared from metal, metal oxide, metal hydroxide or metal carbonate.*
   Write equations for these four possible reactions.
3. *Magnesium can be extracted from magnesium carbonate. Why is the reaction between magnesium and sulphuric acid unlikely to be used to produce magnesium sulphate on a large scale?*

1. sodium sulphate   sodium carbonate   silver nitrate   barium chloride;
2. $Mg(s) + H_2SO_4(aq) \rightarrow MgSO_4(aq) + H_2(g)$; $MgO(s) + H_2SO_4(aq) \rightarrow MgSO_4(aq) + H_2O(l)$; $Mg(OH)_2(s) + H_2SO_4(aq) \rightarrow MgSO_4(aq) + 2H_2O(l)$; $MgCO_3(s) + H_2SO_4(aq) \rightarrow MgSO_4(aq) + H_2O(l) + CO_2(g)$; 3. Magnesium sulphate can be made directly from magnesium carbonate; producing magnesium sulphate directly from magnesium is expensive.

Preparing insoluble salts

KEY POINT **Insoluble salts are prepared by the process of precipitation.**

This involves mixing two solutions, each containing half of the required salt. The salt is then **precipitated**.

Common insoluble salts include silver chloride, barium sulphate, lead sulphate and carbonates, except those of sodium, potassium and ammonium.

e.g. Lead carbonate can be prepared by mixing together the solutions of a lead salt (lead nitrate) and a soluble carbonate (sodium carbonate).

$$Pb(NO_3)_2(aq) + Na_2CO_3(aq) \rightarrow PbCO_3(s) + 2NaNO_3(aq)$$

lead nitrate + sodium carbonate → lead carbonate + sodium nitrate

In order to get a pure sample of lead carbonate, the mixture should be **filtered**, the lead carbonate **washed** with distilled water and **dried**.

KEY POINT **This type of reaction is sometimes called double decomposition and is represented by the equation:**
AX + BY → AY + BX

PROGRESS CHECK

1. *Pick two substances from the list that could be used to prepare lead sulphate by precipitation:*
   lead nitrate **lead carbonate** **sodium sulphate** **sodium carbonate**
2. *Write an ionic equation for the reaction producing lead carbonate.*

Write down the name of the insoluble salt precipitated when each of the following pairs of solutions are mixed:

3. *Sodium sulphate and barium chloride*
4. *Silver nitrate and sodium bromide*
5. *Sulphuric acid and lead ethanoate*
6. *Magnesium sulphate and sodium carbonate*

1. Lead nitrate and sodium sulphate; 2. $Pb^{2+} + CO_3^{2-} \rightarrow PbCO_3$; 3. Barium sulphate;
4. Silver bromide; 5. Lead sulphate;--6. Magnesium carbonate.

11.3 Identification of ions and gases

LEARNING SUMMARY

After studying this section you should be able to:

- ***recall tests for common positive and negative ions***
- ***recall methods used to collect gases and suggest a suitable method to collect a given gas***
- ***recall the tests for some common gases.***

Testing for ions

Testing for metal ions

Testing for ions present is called qualitative analysis.

The presence of many metal ions can be detected by **flame tests**. When compounds of the metal are heated in a hot, blue Bunsen flame – they give the flame a characteristic colour.

The compound being tested is mixed with a little concentrated hydrochloric acid.

A piece of clean platinum (or Nichrome) wire is dipped into the mixture and then held in the flame. **Fig. 11.3** below shows the orange colour imparted to the flame by sodium compounds.

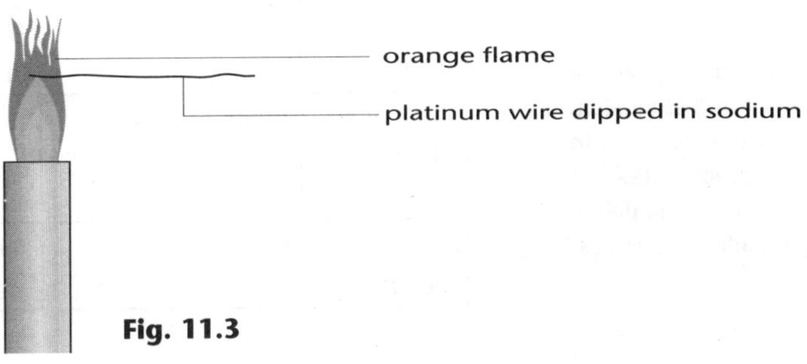

Fig. 11.3

It is not possible to identify many metal ions, e.g. magnesium, as they give the flame no colour.

Table 11.3 below shows the characteristic colours of some metal ions.

Metal ion present	Colour
Sodium	*Orange-yellow*
Potassium	*Lilac-pink*
Lithium	*Red*
Calcium	*Brick red*
Copper	*Green*
Lead	*Blue*
Barium	*Pale green*

Testing for the ammonium ion

A common mistake is to write the ammonium ion as NH_3^+.

The ammonium ion, NH_4^+, can be tested for by using sodium hydroxide solution.

Sodium hydroxide solution is added to the suspected ammonium compound and the mixture is heated in a test tube.

If an ammonium compound is present, **ammonia** gas is produced. This has a pungent smell and turns damp **red litmus paper** held in the mouth of the test tube **blue** (**Fig. 11.4** see below).

Sodium hydroxide turns red litmus blue so it is important that the red litmus paper is held in the gas and it does not touch the side of the test tube where sodium hydroxide might have been.

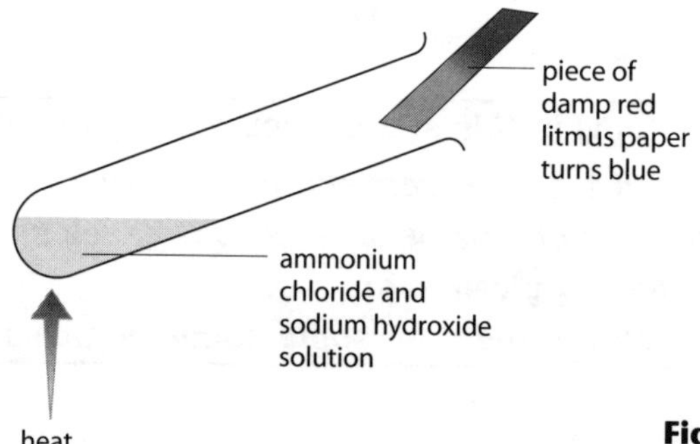

Fig. 11.4

Testing with sodium hydroxide solution

If a small quantity of the compound in solution is tested with **sodium hydroxide solution** an **insoluble hydroxide** may be precipitated. If a precipitate is formed it may redissolve in excess sodium hydroxide solution.

A summary of the precipitation of metal hydroxides with sodium hydroxide solution is shown in **Table 11.4** below.

A similar test can be repeated with ammonia solution as well.

Calcium, magnesium and aluminium ions form white precipitates but only aluminium hydroxide redissolves.

Metal ion	Addition of sodium hydroxide solution		Addition of ammonia solution	
	A couple of drops	*Excess*	*A couple of drops*	*Excess*
Calcium (Ca^{2+})	*White precipitate*	*Precipitate insoluble*	*White precipitate*	*Precipitate insoluble*
Magnesium (Mg^{2+})	*White precipitate*	*Precipitate insoluble*	*White precipitate*	*Precipitate insoluble*
Aluminium (Al^{3+})	*White precipitate*	*Precipitate soluble – colourless solution*	*White precipitate*	*Precipitate insoluble*
Iron (II) (Fe^{2+})	*Green precipitate*	*Precipitate insoluble*	*Green precipitate*	*Precipitate insoluble*

(Table Continued)

(Table Continued)

Metal ion	Addition of sodium hydroxide solution		Addition of ammonia solution	
Iron (III) (Fe^{3+})	*Red brown precipitate*	*Precipitate insoluble*	*Red brown precipitate*	*Precipitate insoluble*
Copper (II) (Cu^{2+})	*Blue precipitate*	*Precipitate insoluble*	*White precipitate*	*Precipitate soluble – deep blue solution*
Zinc (Zn^{2+})	*White precipitate*	*Precipitate soluble – colourless solution*	*White precipitate*	*Precipitate soluble – colourless solution*

Testing for negative ions (anions) in solution

Carbonate

When dilute **hydrochloric acid** is added to a carbonate, **carbon dioxide** gas is produced. No heat is required. The carbon dioxide turns limewater milky.

Sulphate

When dilute hydrochloric acid and **barium chloride** solution are added to a solution of a sulphate, a white precipitate of **barium sulphate** is formed immediately.

Chloride, bromide and iodide

When a solution of chloride, bromide or iodide is acidifed with dilute nitric acid and **silver nitrate** added, a precipitate is formed. A chloride produces a white precipitate of **silver chloride**, a bromide gives a cream precipitate of **silver bromide** and an iodide yields a yellow precipitate of **silver iodide**.

Sulphite

When dilute **hydrochloric acid** is added to a sulphite and the mixture is heated, colourless **sulphur dioxide** is formed. Sulphur dioxide turns a piece of filter paper soaked in potassium dichromate(VI) green.

Nitrate ion (NO_3^-)

If the nitrate is boiled with sodium hydroxide solution and fine aluminium powder (**Devarda's Alloy**), the aluminium powder (a powerful reducing agent) converts the nitrate ion, NO_3^-, into pungent ammonia gas, NH_3.

Methods of collecting gases

The method used to collect a gas depends upon properties of the gas.

One of the huge advances in the eighteenth century was to be able to handle gases, usually by collection over water or over mercury. Today it is important that you can collect gases produced in chemical reactions and identify them with chemical tests.

Gases insoluble or slightly soluble in water

Gases that are insoluble in water or not very soluble in water can be collected **over water** (see **Fig. 11.5**).

Lavoisier collected gases over mercury. This would work with gases that are soluble in water. However, the disadvantages were that the mercury is very toxic and too expensive to use.

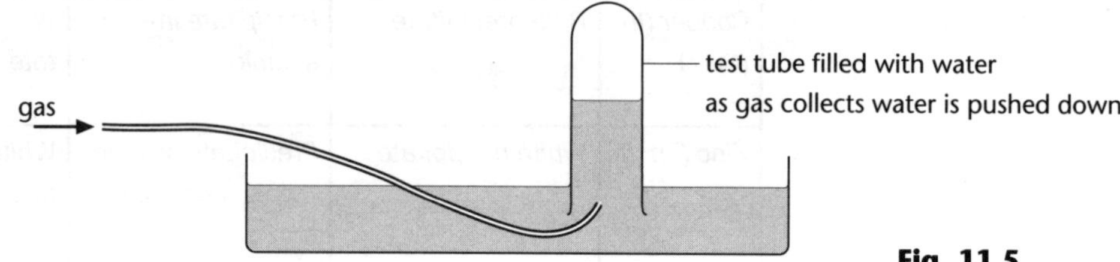

Fig. 11.5

The test tube is filled with water and as the gas is collected the water is pushed out of the tube. If a specific volume of the gas is needed, a measuring cylinder or burette can be used in place of the test tube.

Gases with high and low density

Gases with a **low density** (lighter than air) are collected by **upward delivery** (**Fig. 11.6** below). As the gas is collected, air is pushed out of the tube.

Gases with a **high density** (heavier than air) are collected by **downward delivery** (**Fig. 11.7**). As the gas is collected, air is pushed out of the tube.

All gases can be collected in a gas syringe.

If the gas collected is colourless, it is difficult to ascertain when the tube is full.

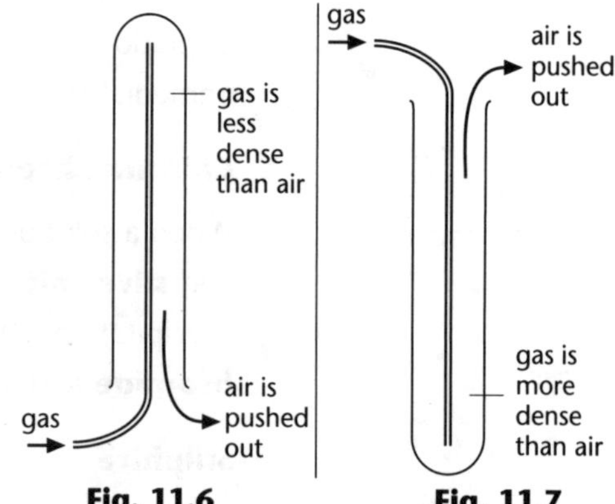

Fig. 11.6 **Fig. 11.7**

PROGRESS CHECK

The table gives some properties of gases labelled A, B, C and D.

Gas	Solubility in water	Density in g/dm³
A	Low	0.08
B	High	0.71
C	High	1.16
D	High	2.66

Density of air under the same conditions 1.21 g/dm³

1. *Which gas could be collected by downward delivery?*
2. *Which gas could be collected by upward delivery?*
3. *Which gas could be collected over water?*
4. *Which gas could not be collected over water or by upward or downward delivery?*

1. D; 2. B; 3. A; 4. C.

Tests for gases

Table 11.5 below gives the tests for common gases.

Gas	Test	Positive result
Oxygen	*Put a* **glowing** *split into gas*	*Splint* **relights**
Hydrogen	*Put a* **lighted** *splint into gas*	**Squeaky pop** *and splint extinguished*
Ammonia	*Add damp* **red litmus paper**	*Turns* **blue**
Carbon dioxide	*Bubble through* **limewater**	*Limewater turns* **milky**
Hydrogen chloride	*Open tube of* **ammonia** *gas and let gases mix*	*Dense* **white fumes** *formed.*
Sulphur dioxide	*Bubble through* **orange potassium dichromate solution**	*Solution turns* **green**
Chlorine	*Add damp* **blue litmus paper**	*Turns* **red** *and then* **bleaches.**

PROGRESS CHECK

1. *Which gas turns litmus paper blue?*
2. *Which gas turns limewater milky?*
3. *A gas turns damp blue litmus paper red and turns potassium dichromate solution green. Which gas is this?*
4. *A gas is mixed with ammonia gas and dense white fumes are formed. What is this gas?*

1. Ammonia; 2. Carbon dioxide; 3. Sulphur dioxide; 4. Hydrogen chloride.

Carbon dioxide gas (CO_2)

When carbon dioxide is passed into limewater (aqueous calcium hydroxide solution), limewater turns cloudy due to the formation of fine milky white precipitate of calcium carbonate.

$$Ca(OH)_2(aq) + CO_2(g) \rightarrow CaCO_3(s) + H_2O(l)$$

Hydrogen gas (H_2)

When a burning splint is brought in contact with hydrogen gas, a pop sound is observed.

$$2H_2(g) + O_2(g) \rightarrow 2H_2O(l) + \text{energy}$$

Oxygen gas (O_2)

When a burning splint is brought in contact with oxygen gas, it re-ignites it into a flame.

$$C(\text{in wood}) + O_2(g) \rightarrow CO_2(g)$$

Ammonia gas (NH_3)

When a damp litmus paper is brought in contact with ammonia gas, it turns red litmus paper blue.

Chlorine gas (Cl_2)

When a damp litmus paper is brought in contact with chlorine gas, it turns blue litmus paper red.

PROGRESS CHECK

1. *What are the flame colours for (a) sodium compounds; (b) potassium compounds (c) calcium compounds?*
2. *What are the colours of (a) copper(II) hydroxide; (b) iron(II) hydroxide; (c) iron(III) hydroxide?*
3. *Which metal hydroxide is insoluble in water but soluble in excess sodium hydroxide solution?*

Use substances in this list to answer questions 4–6.

sodium carbonate	**sodium chloride**
sodium sulphate	**sodium hydroxide**

4. *Which substance in the list fizzes when dilute hydrochloric acid is added and the test tube not heated?*
5. *Which substance in the list forms a white precipitate when dilute nitric acid and silver nitrate are added?*
6. *Which substance in the list produces a gas that turns red litmus blue when heated with ammonium chloride?*

1. (a) Orange-yellow; (b) Lilac-pink ; (c) Brick red; 2. (a) Blue ; (b) Green; (c) Red-brown; 3. Aluminium hydroxide; 4. Sodium carbonate; 5. Sodium chloride; 6. Sodium hydroxide (or sodium carbonate)

Sample IGCSE questions

1. Copper(II) sulphate crystals, $CuSO_4.5H_2O$, can be made by adding excess copper(II) oxide, CuO, which is insoluble in water, to dilute sulphuric acid. The mixture is heated.

(a) Why is it necessary to use excess copper(II) oxide? **[1]**

To make sure all of the acid has reacted ✓.

A common mistake here is to state that solution is saturated because copper(II) oxide remains.

(b) How is the excess copper(II) oxide removed? **[1]**

By filtering ✓.

(c) Describe how copper(II) sulphate crystals are formed from this solution. **[3]**

Evaporate the solution ✓. Until a small volume remains ✓. Allow to cool and crystallise ✓.

It is important that the solution is not evaporated until all the water has gone (evaporated to dryness). Under these conditions crystals would not form and anhydrous copper(II) sulphate would remain.

2. Kerry has four test tubes containing oxygen, hydrogen, carbon dioxide and nitrogen.

The tubes were unlabelled.

Suggest how she could find which gas was which using chemical tests. **[7]**

Put a burning splint into each gas ✓. A squeaky pop will be heard with hydrogen ✓.

A common mistake here is to put in a glowing splint.

Where flame burns brightly — oxygen ✓. The flame extinguished — nitrogen and carbon dioxide ✓.

Test these two with limewater ✓. Limewater turns milky — carbon dioxide ✓ The one which does not turn limewater milky — nitrogen ✓.

It is not enough to put out a flame to confirm carbon dioxide.

Exam practice questions

1. Sue and Sam have three colourless solutions labelled A, B and C.

They know that they are sodium sulphate, sulphuric acid and sodium chloride.

(a) Suggest them a series of tests to identify each chemical. **[6]**

(b) Write an ionic equation for one reaction they used. **[2]**

2. Lead chromate(VI) is an insoluble salt. It is used as a yellow paint pigment.

Describe how a solution of potassium chromate(VI), K_2CrO_4, could be used to produce a dry sample of lead chromate(VI). Write a balanced symbol equation for the reaction. **[8+1]**

(1 mark for Quality of Written Communication)

3. When a mixture of solid ammonium chloride and sodium hydroxide is heated in a test tube ammonia gas is produced.

(a) Finish the symbol equation

$NH_4Cl + NaOH \rightarrow NH_3 + ...$ **[1]**

(b) Ammonia is collected by upward delivery.

(i) Suggest one reason why it is not collected over water. **[1]**

(ii) How could you show when a test tube collecting ammonia is full of ammonia? **[3]**

4.

(a) Given aqueous solutions, 0.2 mol/dm^3, of ammonia and sodium hydroxide, describe how can you show that ammonia is the weaker base. **[2]**

(b) Another compound that contains nitrogen and hydrogen is hydrazine, N_2H_4.

(i) Draw the structural formula of hydrazine. Hydrogen can form only one bond per atom but nitrogen can form three. **[1]**

(ii) Draw a diagram that shows the arrangement of the valence electrons in one molecule of hydrazine. Hydrazine is a covalent compound. Use *x* to represent an electron from a nitrogen atom and *o* to represent an electron from a hydrogen atom. **[3]**

Chapter 12

Metals

The following topics are covered in this section:

- ***Reactivity Series***
- ***Extraction and uses of metals***

12.1 Reactivity series

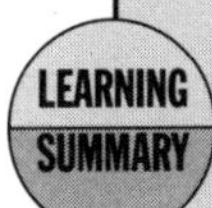

LEARNING SUMMARY

After studying this section you should be able to:

- ***enumerate basic properties of metals***
- ***understand the reactions involving metals on the basis of their position in reactivity series.***

KEY POINT

Metals are used in a wide range of applications. Iron is the metal used in the largest amounts and most of this is in the form of steel. Although steel is useful, it rusts in contact with air and water. Steel is an alloy and alloys are used more than pure metals.

General physical and chemical properties of metals

KEY POINT

Metals:
- **when cut have a typically shiny metallic surface**
- **have comparatively high melting and boiling points**
- **are conductors of electricity**
- **are malleable (i.e. can be hammered into thin sheets) and ductile (i.e. can be drawn into thin wires)**
- **have high density.**

Metals can be arranged in order of reactivity by conducting experiments on various metals in the laboratory and finding out the respective speeds of reaction.

Reaction with acids (dilute hydrochloric acid)

Group 1 elements in the periodic table are the most reactive.

Metals, on reaction with HCl, form metal chloride and hydrogen is released in the process.

E.g. $Mg(s) + 2\ HCl(aq) \longrightarrow MgCl_2(s) + H_2(g)$

The rate of formation of hydrogen increases with the increase in reactivity of the metal.

Metals can be arranged in order of reactivity by measuring hydrogen gas, collected in a given time during reaction with HCl.

Reaction with water or steam

Sodium and potassium are kept in kerosene since they react violently with air and water.

Metals, on reaction with water/steam, form metal oxide and hydrogen is released in the process.

E.g. Sodium reacts with water and magnesium reacts with steam as follows:

$$2\,Na(s) + H_2O(l) \longrightarrow 2\,NaOH(aq) + H_2(g)$$

$$Mg(s) + H_2O(g) \longrightarrow MgO(s) + H_2(g)$$

The rate of formation of hydrogen increases with the increase in reactivity of the metal.

Unreactive metals are more useful in daily life.

Metals can be arranged in order of reactivity by measuring the volume of hydrogen gas collected in a given time during reaction with H_2O.

Table 12.1 below gives a list of metals arranged in order of reactivity.

Periodic group	Metal	Formula	Reaction with dilute HCl	Reaction with water/ steam	Reduction of oxide with carbon	Ease of extraction from ores
1 (Alkali Metal)	*Potassium*	*K*	*Very high*	*Very high, burn violently in water*	*None*	*Difficult to extract*
	Sodium	*Na*	*Very high*			*Difficult to extract*
2 (Alkaline Earth metal)	*Calcium*	*Ca*	*High*	*Medium (with Steam)*	*None*	*Easier to extract*
	Magnesium	*Mg*	*High*	*Medium (with Steam)*	*None*	*Easier to Extract* ↓
3 Transition Metals	*Aluminium*	*Al*	*Medium*	*Low (with steam)*	*None*	
	Zinc	*Zn*	*Medium*	*Low (with steam)*	*Metal oxides reduced by carbon to yield pure metal* ↓	
	Iron	*Fe*	*Medium*	*Low (with steam)*		
4 Transition Metals	*Lead*	*Pb*	*Low*	*None*		
	Copper	*Cu*	*None*	*None*		
	Silver	*Ag*	*None*	*None*		*Found in native state*
	Gold	*Au*	*None*	*None*		*Found in native state*
	Platinum	*Pt*	*None*	*None*		*Found in native state*

Reactivity - tendency of a metal to form its positive ion

- **The more reactive a metal, the greater is its tendency to lose electrons and form a positive ion.**
- **A more reactive metal, on reaction with salts of less reactive metals, transfers electrons to the less reactive metal ion and displaces the less reactive metal ion to form its own salt. The less reactive metal gets isolated to its elemental state in the process.**

Reduction of metal oxides

A more reactive metal, on reaction with oxides of less reactive metals, takes away the oxygen and forms its own oxide.

During reaction, both the metals compete for the oxygen atom and the more reactive metal *snatches* it away after a violent process.

The less reactive metal is reduced to its elemental state in the process.

E.g. $$Fe_2O_3(s) + 2\,Al(s) \xrightarrow{\text{Heat}} Al_2O_3\,(s) + 2Fe\,(s)$$

Similarly, carbon a non metal, due to its higher reactivity can reduce the oxides of metals below aluminium in **Table 6.1**, to yield pure metals.

Reactions in aqueous solutions

In aqueous solutions, a more reactive metal displaces a less reactive metal from the solution of its salt.

The less reactive metal is reduced to its elemental state in the process.

This type of reaction is known as displacement reaction.

E.g. Zn, a more reactive metal than Cu, reacts with copper nitrate in aqueous solution as follows:

$$Cu(NO_3)_2(aq) + Zn(s) \longrightarrow Zn(NO_3)_2\,(aq) + Cu\,(s).$$

Action of heat on metal compounds

Metal hydroxides, when heated, are converted to their oxides and water is released in the process.

E.g. $$Na(OH)_2 \xrightarrow{\text{Heat}} NaO + H_2O$$

Hydroxides of metals such as iron, lead and copper, when heated, change colour as a result of oxide formation.

Metal nitrates, when heated, are converted to their oxides and nitrogen dioxide (a poisonous gas) is released in the process.

E.g. $$NaNO_3 \xrightarrow{\text{Heat}} NaO + NO_2$$

12.2 Extraction and uses of metals

LEARNING SUMMARY

After studying this section you should be able to:

- ***understand how metals can be extracted from metal ores***
- ***explain why aluminium seems less reactive than would be expected from its position in the reactivity series***
- ***understand the chemical principles involved in the extraction of metals like iron by reduction***
- ***explain the extraction of zinc by froth flotation process***
- ***recall the names, composition and uses of alloys.***

Extracting metals from ores

The method used to extract the metal from the ore depends on the position of the metal in the reactivity series.

If a metal is high in the reactivity series its ores are stable and the metal can be obtained only by electrolysis.

Metals that are obtained by electrolysis include potassium, sodium, calcium, magnesium and aluminium.

You should be able to predict the method used to extract a metal from its ores, given its position in the reactivity series.

Metals in the middle of the reactivity series do not form very stable ores and they can be extracted by reduction reactions, often with carbon.

Examples of metals extracted by reduction are zinc, iron and lead.

Metals occuring at lower positions in the reactivity series, if present in ores, can be extracted simply by heating because the ores are

For example, mercury can be extracted by heating *cinnabar*. A few metals such as gold are found uncombined in the Earth.

Reactivity of aluminium

Aluminium is below magnesium but above zinc in the reactivity series.

It would be expected that aluminium is less reactive than magnesium but more reactive than zinc.

Table 12.2 below shows that this is not actually observed.

Potassium
Sodium
Calcium
Magnesium
Aluminium
Zinc
Iron

Metal	Reaction with dilute hydrochloric acid
Magnesium	*Forms bubbles of gas steadily*
Aluminium	*No reaction after 10 minutes*
Zinc	*Slow reaction*

Aluminium reacts much less than would be expected from its position in the reactivity series. The surface of aluminium is coated with a thin layer of aluminium oxide. Aluminium oxide is very unreactive. This unreactive aluminium oxide layer keeps reactants away from the aluminium and hence prevents reaction.

Aluminium has a dull appearance. This is due to the oxide layer. If this layer is removed, aluminium becomes more reactive. A piece of aluminium foil heated in a Bunsen flame does not burn. Dipping aluminium foil into mercury removes the oxide layer. Aluminium foil then reacts on standing in the air to form aluminium oxide.

Anodising aluminium

As previously stated, aluminium is coated with a thin layer of aluminium oxide.

The process of anodising is used to thicken the layer by electrolysis to give further protection.

This oxide layer can also be coloured by dyes absorbed into the layer to give a decorative finish.

Fig. 12.1 below shows apparatus suitable for anodising a piece of aluminium. Aluminium is used as the anode of the cell. Dilute sulphuric acid is the electrolyte. During electrolysis, oxygen is produced at the anode and this reacts with the aluminium.

$$2O^{2-} \rightarrow O_2 + 4e^-$$

$$4Al(s) + 3O_2(g) \rightarrow 2Al_2O_3(s) \text{ (overall).}$$

Remember that anodising takes place at the anode so the piece of aluminium is made the anode.

Anodised aluminium is used for making window frames, racing yachts and windsurfing boards.

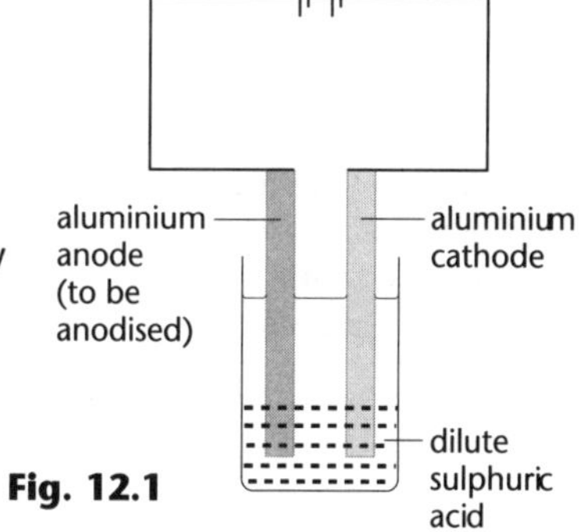

Fig. 12.1

PROGRESS CHECK

1. *What is formed as a thin layer on the surface of aluminium?*
2. *Why does this make the aluminium less reactive than expected?*
3. *Aluminium placed in dilute sulphuric acid does not react for a long period of time. Eventually after half-an-hour the reaction starts and bubbles of gas are collected. Explain why there is this delay.*
4. *During anodising, which ion migrates towards the cathode?*
5. *What is produced at the cathode?*
6. *Write an ionic equation for the discharging of the ions at the cathode.*

1. Aluminium oxide; 2. It prevents the reactant coming into contact with the aluminium surface; 3. It takes a long time for the aluminium oxide surface coating to break down and then reaction can take place; 4. Hydrogen ions, H^+; 5. Hydrogen; 6. $2H^+ + 2e^- \rightarrow H_2$

Extracting metals by reduction

Iron is an example of a metal extracted by **reduction**. The reducing agent is **carbon monoxide**. This removes oxygen from the iron oxide to leave iron.

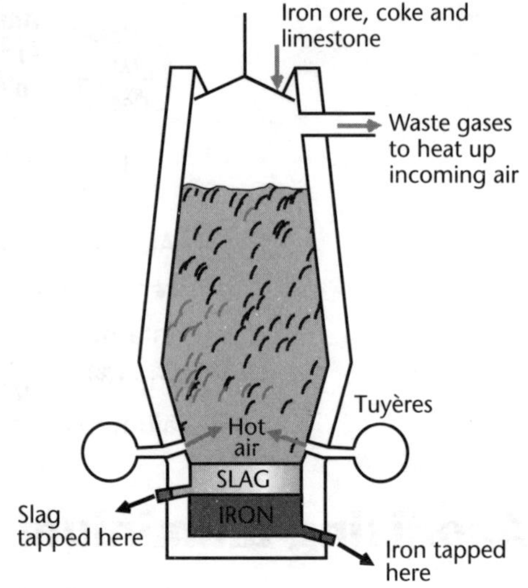

Fig. 12.2 Blast furnace

The extraction of iron is carried out in a **blast furnace** (**Fig. 12.2**).

The furnace is loaded with **iron ore**, **coke** and **limestone** and is heated by blowing hot air into the base from the tuyères. This raises the temperature to about 1500°C. Inside the furnace the following reactions take place:

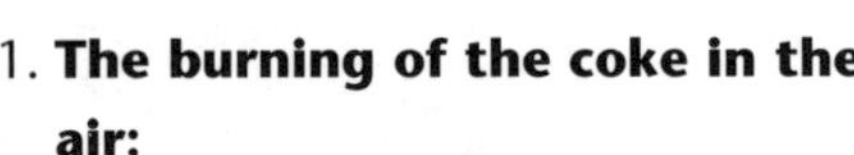

1. **The burning of the coke in the air:**

   $C(s) + O_2(g) \rightarrow CO_2(g)$

   carbon + oxygen → carbon dioxide

2. **The reduction of the carbon dioxide to carbon monoxide:**

   $CO_2(g) + C(s) \rightarrow 2CO(g)$

   carbon dioxide + carbon → carbon monoxide

This is the important reduction step.

3. **The reduction of the iron ore to iron by carbon monoxide:**

   $Fe_2O_3(s) + 3CO(g) \rightarrow 2Fe(l) + 3CO_2(g)$

   iron(III) oxide + carbon monoxide → iron + carbon dioxide

4. **The decomposition of the limestone produces extra carbon dioxide:**

   $CaCO_3(s) \rightarrow CaO(s) + CO_2(g)$

   calcium carbonate → calcium oxide + carbon dioxide

This step removes impurities from the furnace so it can keep working.

5. **The removal of impurities by the formation of slag:**

   $CaO(s) + SiO_2(s) \rightarrow CaSiO_3(l)$

   calcium oxide + silicon dioxide → calcium silicate (slag)

The molten iron sinks to the bottom of the furnace and the slag floats on the surface of the molten iron. Periodically, the **iron** and **slag** can be tapped off.

The iron produced is called **pig iron** and contains about 4 per cent carbon. Most of this is turned into the alloy called **steel**.

The slag is used as a **phosphorus fertiliser** and for **road building**.

Extraction of Zinc

KEY POINT

Principal ore of zinc is zinc sulphide or zinc blende.

Zinc is a fairly reactive metal. The main process of extraction of Zinc from its ore involves the following steps:

- zinc ore is first concentrated by **froth floation**.
- The concentrated zinc sulphide is then heated in a furnace to form zinc oxide.
- The zinc oxide is mixed with powdered coke(carbon) and heated strongly (to approx 1400° C).
- Carbon being more reactive, reduces the zinc oxide to zinc.

The following reactions take place

$$2\,ZnS(s) + 3\,O_2(g) + \xrightarrow{\text{Heat}} 2\,ZnO(s) + 2SO_2(g)$$

$$ZnO(s) + C(s) + \xrightarrow{\text{Heat}} Zn(s) + CO(g)$$

Uses of Zinc

- Zinc is used in manufacture of alloys, e.g. brass.
- It is also used for galvanising steel.
- In batteries zinc is used as an electrode.

Alloys

Alloys are mixtures of metals or mixtures of metals with carbon. Alloys have many applications because they have better properties for many uses than pure metals.

Table 12.3 below gives the uses of some pure metals.

Metal	Use	Reason for use
Copper	*Copper wires*	*Excellent conductor of electricity/very ductile*
Tin coating	*tin cans*	*Not poisonous*
Aluminium	*Kitchen foil*	*Very malleable*
Iron	*Wrought iron gates*	*Easy to forge*
Lead	*Flashing on roofs*	*Soft, easy to shape/very unreactive*

An alloy can be made by weighing out correctly the different constituent metals and then melting them together.

Fig. 12.3 shows the arrangement of particles in a pure metal and in an alloy.

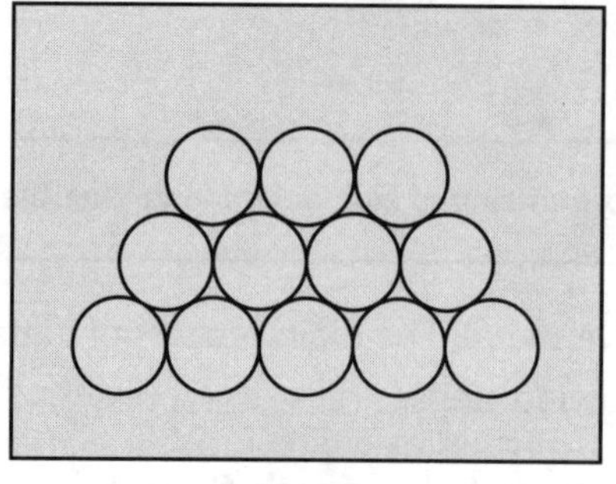
pure metal

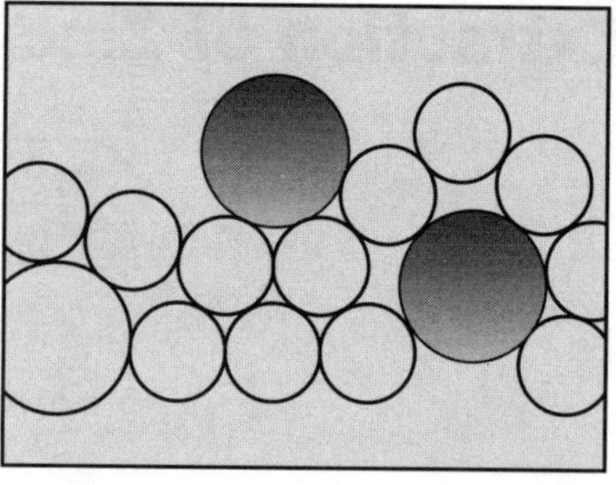
alloy

Fig. 12.3

KEY POINT **The particles in a pure metal are able to move past each other much better than the particles in an alloy. This makes the alloy harder.**

Table 12.4 below gives the composition and uses of some common alloys.

Alloy	Constituent elements	Uses
Steel	*Iron + between 0.15% and 1.5% carbon. The properties of steel depend upon the percentage of carbon.*	*Wide variety of uses including cars,ships, tools, reinforced concrete*
Brass	*Copper and zinc*	*Ornaments , buttons, screws*
Duralumin	*Aluminium, magnesium, copper and tin*	*Lightweight uses e.g. aircraft panels, bicycle frames*
Solder	*Tin and lead*	*Joining metals together*
Bronze	*Copper and tin*	*Statues*

Steel

KEY POINT **Most of the iron produced in a blast furnace is turned into steel. The iron from the blast furnace, called pig iron, contains impurities of carbon, phosphorus and silicon.**

The steel-making process involves removing all of the impurities to get pure iron and then adding the correct amounts of different materials to get steel of the required quality.

The steel-making furnace (**Fig. 12.4**) is tilted and loaded with 30% scrap iron and 70% molten iron from the blast furnace.

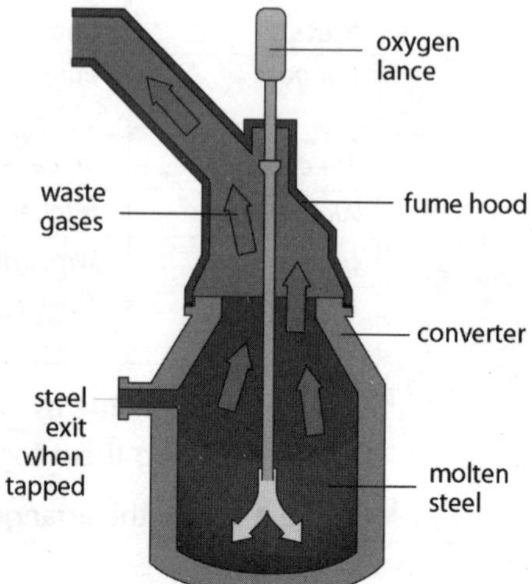

Fig. 12.4

A water-cooled lance is lowered into the upright furnace and pure **oxygen** is blown, under high pressure, onto the surface of the molten iron. The oxides of carbon and phosphorus escape as gases. **Limestone** is added to remove the other impurities as slag. Finally, the required amounts of carbon and other elements are added to give steel of the required quality.

Table 12.5 compares the properties and uses of three types of steel.

Type of steel	Properties of steel	Percentage of carbon
Low carbon steel	*Soft and easily shaped*	*0.03–0.25*
Mild steel	*Easily pressed into shape*	*0.25–0.50*
High carbon steel	*Strong but brittle*	*0.85–1.50*

Stainless steel contains transition metals such as chromium and nickel. It does not rust easily as do iron and other types of steel.

Uses of Steel

- Steel is the most widely used metal on earth.
- Various types of steel have different uses.

Mild steel being ductile and easily workable is used in manufacture of bodies of automobiles and machinery.

Stainless steel being non corrosive is widely used as cutlery, surgical instruments and in reaction vessels of chemical plants

1. *Which type of steel is most suitable for making drill bits?*
2. *Why is it better to remove all of the impurities from steel during manufacture and then add the carbon again afterwards?*
3. *Solder is an alloy of tin and lead. It has a lower melting point than tin or lead. Why is this an advantage when soldering metals together?*
4. *Why is duralumin better for constructing aeroplanes but pure aluminium better for overhead power cables?*

1. High carbon steel; 2. By removing all impurities it is easier to get the correct amounts in the steel at the end of the process; 3. When metals are soldered the solder has to be melted. A low melting point means it is easier to melt the solder; 4. Duralumin is stronger and this is needed for aircraft structure, but pure aluminium is a better conductor of electricity.

Sample IGCSE questions

1. Different methods are used to extract iron and aluminium from their ores.

(a) What is an ore? **[2]**

A mineral that has enough metal (or metal compound) ✓ in it to make it worthwhile extracting the metal ✓.

You may be allowed 'a rock' instead of 'a mineral'.

(b) **(i)** Write down the raw materials used to extract iron from iron oxide in the blast furnace. **[3]**

Coke ✓, limestone ✓ and air ✓

Don't count 'iron ore', you are given this. Oxygen may be allowed instead of air.

(ii) What name do we give to the removal of oxygen from a metal oxide? **[1]**

Reduction ✓

(c) Aluminium is extracted from aluminium oxide by passing electricity through molten aluminium oxide dissolved in cryolite.

(i) What is the job of the cryolite? **[2]**

Its job is to dissolve the aluminium oxide at a much lower temperature than the melting point of aluminium oxide ✓. Cryolite has a lower melting point than aluminium oxide ✓.

(ii) Explain why the aluminium oxide has to be molten or dissolved in the molten cryolite for the extraction process to work. **[1]**

This is electrolysis and, for electrolysis to work, the ions have to be free to move about. The ions are free to move about in a molten substance or when a substance is dissolved. They then can move to oppositely charged electrodes ✓.

Do not make a vague comment such as 'to make it cheaper'. A long answer for one mark, but it ensures the question is answered.

(iii) Finish the overall equation for the extraction of aluminium. **[2]**

$Al_2O_3 \rightarrow$ ……… + ………….

$2Al_2O_3 \rightarrow 4Al + 3O_2$ ✓✓

One mark is for the correct products and one mark for balancing.

(iv) Write down the name of the gas formed at the positive electrode and explain why the positive electrode has to be replaced regularly . **[2]**

Oxygen gas is formed ✓ and this burns the carbon electrodes away very quickly ✓.

Remember the carbon electrodes burn away to form carbon dioxide gas.

Sample IGCSE questions

(d) Suggest two properties of aluminium that make it suitable for use in overhead power cables. **[2]**

It has low density ✓ and is a good conductor of electricity ✓.

Do not write 'it is light', put 'low density'. You may be allowed 'resists corrosion' but do not give answers such as 'shiny'.

(e) Today, much smaller amounts of iron and aluminium are used than 50 years ago. Suggest reasons why is this so. **[3]**

New materials have been developed, e.g. polymers and composites ✓. Much thinner sheets of metals are used ✓. More metals are recycled ✓.

There may be other possible answers but you should give at least three because there are three marks.

Exam practice questions

1. Steel is the most widely used alloy.

Describe how steel is formed from iron ore. **[4]**

2. Zinc blende is the common ore of zinc. It is usually found mixed with an ore of lead and traces of silver.

(i) Describe how zinc blende is changed into zinc oxide. **[2]**

(ii) Write an equation for the reduction of zinc oxide by carbon. **[2]**

(iii) The boiling point of lead is 1740 °C and that of zinc is 907 °C. When both oxides are reduced by heating with carbon at 1400 °C, only lead remains in the furnace. Explain. **[2]**

3.(a) A major use of zinc is to make diecasting alloys. These alloys contain close to 4% of aluminium, they are less malleable than pure zinc but are stronger.

(i) Give one other industrial use of aluminium (Al). **[1]**

(ii) Describe the structure of a typical metal, such as zinc, and explain why it is malleable. **[3]**

(iii) Suggest why the introduction of a different metallic atom into the structure makes the alloy stronger than the pure metal. **[2]**

(b) A solution of an impure zinc ore contained zinc, lead and silver(I) ions. The addition of zinc dust will displace both lead and silver.

(i) The ionic equation for the displacement of lead is as follows:

$$Zn(s) + Pb^{2+}(aq) \rightarrow Zn^{2+}(aq) + Pb(s)$$

Change 1 happens for zinc and change 2 happens for lead.

Which one of the two changes is reduction? Explain your answer. **[2]**

(ii) Write an ionic equation for the reaction between zinc atoms and silver(I) ions. **[2]**

Chapter 13

Electrochemistry and electrolysis

The following topics are covered in this section:

- *Cells and conductivity*
- *Some examples of electrolysis*
- *Uses of electrolysis*

Electrolysis is an important process. It was developed by Michael Faraday in the early nineteenth century when suitable supplies of electricity became available. He discovered a number of new elements, e.g. the alkali metals and established the basic laws of electrolysis.

13.1 Cells and conductivity

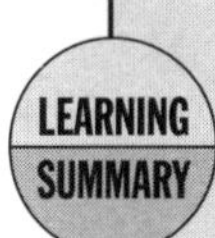

After studying this section you should be able to:

- ***recall that metals and graphite conduct electricity (when solid) because they contain free moving electrons***
- ***recall that electrolytes such as sodium chloride conduct electricity and are decomposed when molten or in solution***
- ***describe how a fuel cell can be made and the advantages of fuel cells.***

Electrical conductivity

Fig. 13.1 below shows apparatus that can be used to test if a substance conducts electricity. If electricity passes through X, the bulb lights.

The only common examples of solids conducting electricity are metals and graphite (a form of carbon).
Electricity passes through metals and graphite because of freely moving (delocalised) electrons.
Electricity passes through metals and graphite with no chemical change taking place.

Liquid metals also conduct electricity and again no decomposition takes place.

Solid acids, alkalis and salts (called **electrolytes**) do not conduct electricity. Although they contain ions, the ions are held in fixed positions and are not able to move.

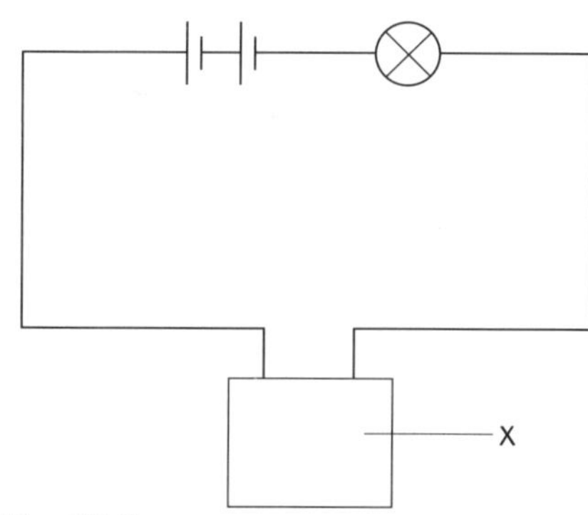

Fig. 13.1

Melting an electrolyte or dissolving it in water breaks up the structure and the ions are free to move.
Electrolysis is the splitting up of an electrolyte when molten or in solution.

Electrolytes contain **ionic bonds**.

Ions move (or migrate) towards the electrode of opposite charge.

This is shown in **Fig. 13.2.**

When the ions reach the electrode they may be **discharged**. This involves either a transfer of electrons to or from the electrode.

Compounds that are liquid or in solution but do not conduct electricity contain **covalent bonds**.

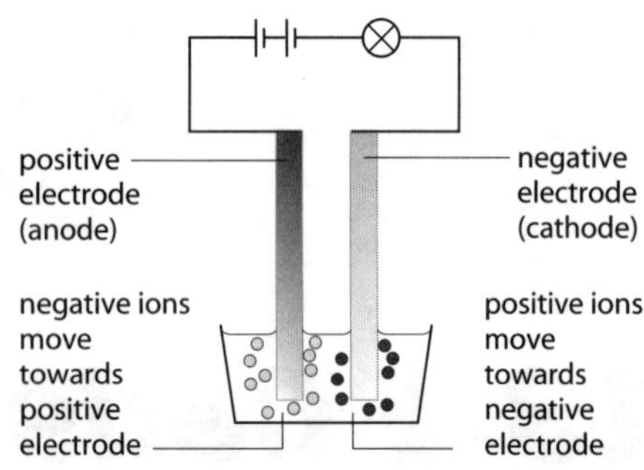

Fig. 13.2

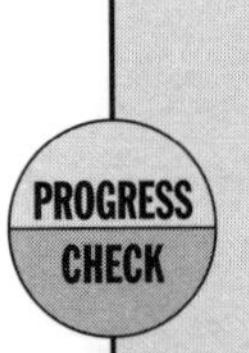

copper **ethanol** **graphite** **mercury**
molten sodium chloride **poly(ethene)** **sodium**
sodium chloride solution **solid sodium chloride**

1. *Which of the substances in the list will conduct electricity?*
2. *Which of the substances in the list will conduct electricity without decomposition?*

1. Copper, graphite, mercury, molten sodium chloride, sodium, sodium chloride solution;
2. Copper, graphite, mercury, sodium

Cells

Electricity can be produced by cells where a chemical reaction is taking place.

Fig. 13.3 below shows two pieces of copper dipping in an **electrolyte** (solution of sodium chloride).

The two pieces of copper are connected together with a copper wire and light bulb.

Fig. 13.3

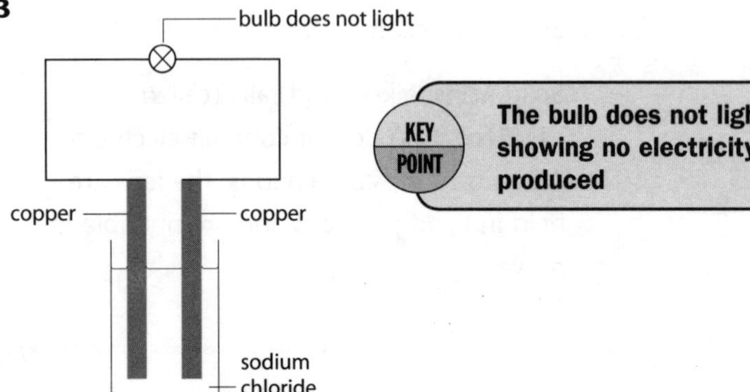

The bulb does not light up, showing no electricity is produced

Fig. 13.4 shows **two different metals** – copper and magnesium dipped in sodium chloride solution.

In any cell a chemical reaction is taking place. The energy is released as electrical energy. The cell stops working when one of the chemicals is used up.

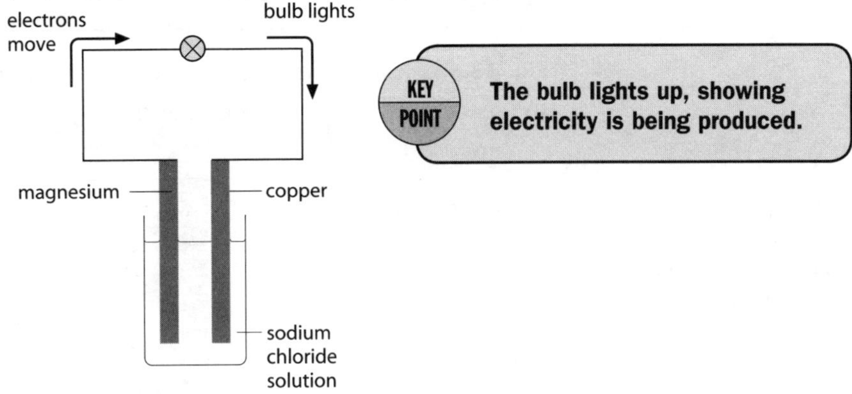

Fig. 13.4

> **KEY POINT** The bulb lights up, showing electricity is being produced.

Electrons are moving through the wire from the magnesium rod to the copper rod. If a voltmeter is put in the wire instead of light bulb a voltage of 2.7 volts can be measured. This is a measure of the pushing power in the circuit.

At the magnesium, electrons are lost.

$$Mg \rightarrow Mg^{2+} + 2e^-$$

Magnesium is the **negative pole** of the cell and copper is the **positive pole** of the cell.

> **KEY POINT** A cell consists of two different metals dipping in a solution that conducts electricity (electrolyte). In the cell, chemical energy is converted into electrical energy. The more reactive metal becomes the negative pole of the cell from which electrons flow.

Using different metals in cells

Table 13.1 below shows the voltages in different cells. In each cell, one metal (rod A) is copper and the other metal (rod B) is changed. The same salt solution is used in each case.

Rod A	Rod B	Voltage (V)
Copper	*Zinc*	*0.60*
Copper	*Iron*	*0.30*
Copper	*Lead*	*0.02*
Copper	*Copper*	*0.00*
Copper	*Silver*	*-0.5*

> **KEY POINT** The further apart the metals are in the reactivity series the higher will be the voltage of the cell.

Fuel cells

KEY POINT **A fuel cell is a special type of cell which converts chemical energy into electrical energy in an efficient way.**

Fig. 13.5 below shows a diagram of a fuel cell that produces electricity using the reaction between hydrogen and oxygen.

A hydrogen-oxygen fuel cell does not produce any pollution. The only product is water.

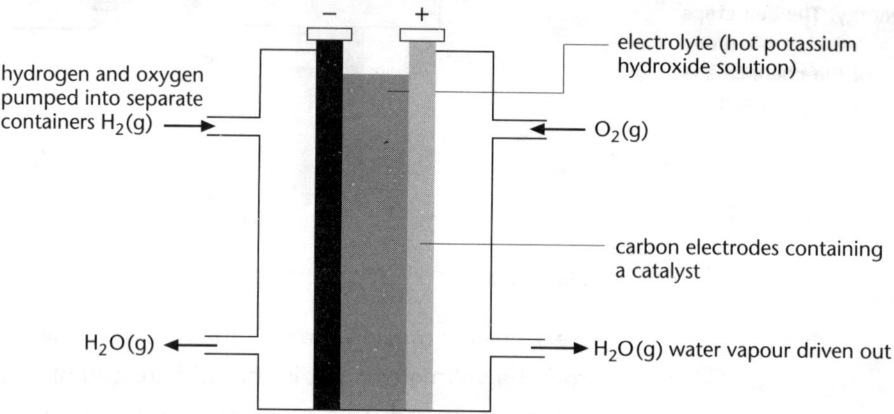

Fig. 13.5

Reactions in the hydrogen–oxygen fuel cell

At the positive electrode:

$$2H_2(g) + 4OH^-(aq) \rightarrow 4H_2O(l) + 4e^-$$

At the negative electrode:

$$O_2(g) + 2H_2O(l) + 4e^- \rightarrow 4OH^-(aq)$$

Overall reaction:

$$2H_2(g) + O_2(g) \rightarrow 2H_2O(l)$$

For fuel cells to be widely used a cheap source of hydrogen is needed.

KEY POINT **This is an exothermic reaction and all of the energy produced is produced as electricity.**

Fuel cells are used in space vehicles and they are also being tried in cars instead of petrol engines.

PROGRESS CHECK

1. *A cell is set up with two rods dipping into a solution. Which of the sets of apparatus would produce a voltage on a voltmeter?*

	Rod S	Rod T	Solution
A	*zinc*	*lead*	*glucose*
B	*zinc*	*zinc*	*sodium sulphate*
C	*zinc*	*lead*	*sodium sulphate*
D	*lead*	*lead*	*sodium sulphate*

2. *A simple cell was set up with copper and silver dipping into sodium chloride solution. Which metal is the negative electrode?*
3. *What is produced in a hydrogen–oxygen fuel cell?*
4. *Why is an engine burning hydrogen and oxygen better than a petrol-burning engine?*
5. *Why is a fuel cell better than an engine burning hydrogen?*

1. C; 2. Copper (higher in the reactivity series); 3. Water; 4. The only product is water which does not pollute the atmosphere – no carbon dioxide, carbon monoxide or sulphur dioxide; 5. All of the energy is produced as electricity and none is lost as heat to the surroundings.

13.2 Some examples of electrolysis

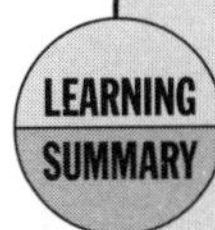

After studying this section you should be able to:

- ***explain what is happening during the electrolysis of molten lead bromide***
- ***explain the products of electrolysis of some aqueous solutions of electrolytes and be able to predict others.***

Electrolysis of molten lead(II) bromide

The apparatus in **Fig. 13.6** could be used for the electrolysis of lead bromide, $PbBr_2$:

Fig. 13.6

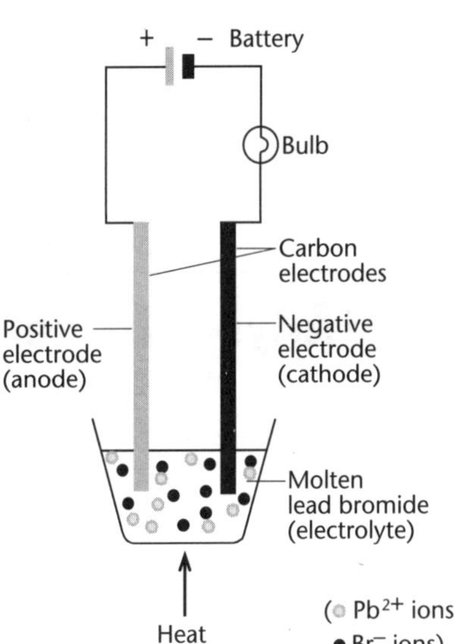

- the bulb does not light while the lead bromide is solid
- as soon as the lead bromide melts, the bulb lights
- lead starts to form around the negative electrode
- bromine gas is produced at the positive electrode
- the bulb goes out when heat is removed and the melt solidifies.

The overall reaction is

$$PbBr_2 \rightarrow Pb + Br_2$$

The carbon rods are called **electrodes**.

The electrode attached to the positive terminal of the battery is the positive electrode (sometimes called the **anode**).

The electrode attached to the negative terminal of the battery is the negative electrode (sometimes called the **cathode**).

The positive electrode has a shortage of electrons and the negative electrode has a surplus of electrons. Electrons are constantly flowing through the wire from the positive electrode to the negative electrode.

Migration of ions

Ions move towards the electrode of opposite charge.

When lead bromide is melted, **positive** lead ions, Pb^{2+}, move towards the **negative** electrode.

The **negative** chloride ions, Br^-, move towards the positive electrode.

Discharging of ions

Remember the cathode has an excess of electrons. This process involves a gain of electrons and so is an example of reduction.

When the ions arrive at the electrodes they are **discharged**.

At the negative electrode (cathode) electrons are transferred from the cathode to the ion and the ion is changed to a metal atom.

$$Pb^{2+} + 2e^- \rightarrow Pb$$

Remember the anode has a shortage of electrons and this change is providing more electrons. This process involves a loss of electrons and so is an example of oxidation.

KEY POINT **At the positive electrode (anode), electrons are transferred from the ion to the anode and the ion is changed to a bromine atom. Two bromine atoms combine to form a bromine molecule.**

$Br^- \rightarrow Br + e^-$ $\quad$ $2Br \rightarrow Br_2$

PROGRESS CHECK

1. *In an experiment to electrolyse molten lead bromine the electrolysis was carried out for a long time. On cooling, the bulb continued to glow brightly. Explain why this might happen.*

A similar experiment was carried out with lithium iodide, LiI, used in place of lead bromide.

2. *Which ions are present in lithium iodide?*
3. *Which ion moves towards the cathode?*
4. *Which ion moves towards the anode?*
5. *What is produced at the cathode?*
6. *What is produced at the anode?*
7. *Write an ionic equation for the change at the cathode.*
8. *Write an ionic equation for the change at the anode.*

1. The lead produced links the anode and cathode and electricity passes through it without going through the melt; 2. Li^+ and I^-; 3. Li^+; 4. I^-; 5. Lithium; 6. Iodine;
7. $Li^+ + e^- \rightarrow Li$; $I^- \rightarrow I + e^-$ and $I + I \rightarrow I_2$ or $2I^- \rightarrow I_2 + 2e^-$

Electrolysis of aqueous solutions

In pure water, about 1 in every 600 000 000 water molecules is split into hydrogen and hydroxide ions.

$$H_2O(l) \rightleftharpoons H^+(aq) + OH^-(aq)$$

This slight ionisation of water explains why pure water has a low electrical conductivity.

Electrolysis of sodium chloride solution

KEY POINT **A solution of sodium chloride in water contains the following ions:**

$H^+(aq)$	**$OH^-(aq)$**	**from the water**
$Na^+(aq)$	**$Cl^-(aq)$**	**from sodium chloride.**

Both positive ions migrate to the cathode (negative electrode) and both negative ions migrate to the anode (positive electrode). At the electrode, either one or both of the ions are discharged.

Table 13.2 below gives the results of some electrolysis experiments

Solution	Electrodes	Ion discharged at positive electrode	Product at at positive electrode	Ion discharged at negative electrode	Product at negative electrode
Dilute sulphuric acid	*carbon*	*$OH^-(aq)$*	*oxygen*	*$H^+(aq)$*	*hydrogen*
Dilute sodium hydroxide	*carbon*	*$OH^-(aq)$*	*oxygen*	*$H^+(aq)$*	*hydrogen*
Copper(II) sulphate	*carbon*	*$OH^-(aq)$*	*oxygen*	*Cu^{2+} (aq)*	*copper*
Copper(II) sulphate	*copper*	*none*	*none*	*Cu^{2+} (aq)*	*copper*
Copper(II) chloride	*carbon*	*$Cl^-(aq)$*	*chlorine*	*Cu^{2+} (aq)*	*copper*
Very dilute sodium chloride	*carbon*	*$OH^-(aq)$*	*oxygen*	*$H^+(aq)$*	*hydrogen*
Concentrated sodium chloride	*carbon*	*$Cl^-(aq)$*	*chlorine*	*$H^+(aq)$*	*hydrogen*
Concentrated sodium chloride	*mercury cathode*	*$Cl^-(aq)$*	*chlorine*	*$Na^+(aq)$*	*sodium amalgam*

The apparatus in **Fig. 13.7** can be used for the electrolysis of aqueous solutions. It enables the gases produced at the electrodes to be collected.

- **Metals**, if produced, are produced at the **negative** electrode.
- **Hydrogen** is produced at the **negative** electrode only.
- **Non-metals**, apart from hydrogen, are produced at the **positive** electrode.
- **Reactive metals** are not formed at the positive electrode during the electrolysis of aqueous solutions. An exception occurs during the electrolysis of sodium chloride using a mercury cathode.

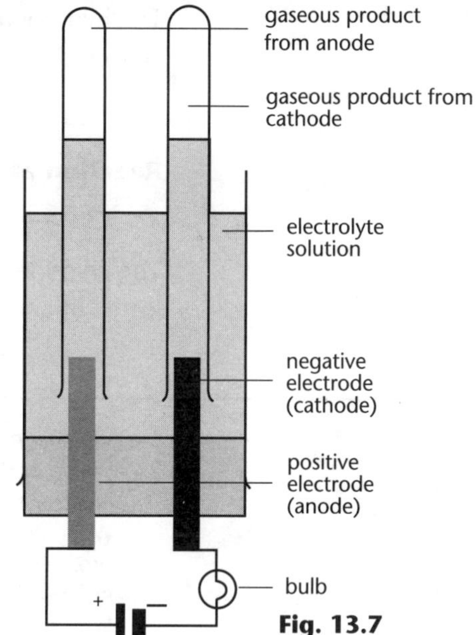

Fig. 13.7

- The products can depend upon the concentration of the electrolyte in the solution. For example, electrolysis of concentrated sodium chloride solution produces chlorine at the positive electrode but the electrolysis of dilute sodium chloride can produce oxygen.
- Providing the concentrations of the negative ions in solution are approximately the same, the order of discharge is

$OH^-(aq)$
$I^-(aq)$
$Br^-(aq)$
$Cl^-(aq)$
$NO_3^-(aq)$
$SO_4^{2-}(aq)$

↓ ease of discharge decreases

Electrolysis of acids

In aqueous solutions, the acids ionise to give the cations (H^+ ions) and the anions derived from the non metals.
E.g.

$$HNO_3 \rightleftharpoons H^+ + NO^-_3$$
$$H_2SO_4 \rightleftharpoons 2H^+ + SO^{2-}_4$$

Electrolysis of concentrated hydrocloric acid

The aqueous solutions of HCl undergoes electrolysis as follows:

Ionisation : $HCl \rightleftharpoons H^+ + Cl^-$

H^+ ions move towards cathode (Negative electrode) and Cl^- ions move towards anode (positive electrode).

Reaction at the cathode

$$H^+ + e^- \text{ (Receved from cathode)} \longrightarrow H \text{ (Primary reaction)}$$

The hydrogen atoms thus formed further react to form hydrogen gas, which is collected at the cathode.

$$H + H \longrightarrow H_2 \text{ (Secondary reaction)}$$

Reaction at the anode

$$Cl^- \longrightarrow Cl + e^- \text{ (Collected by the anode)} \quad \text{(Primary reaction)}$$

The chlorine atoms thus formed, combine to form chlorine gas which is collected at anode.

$$Cl + Cl \longrightarrow Cl_2 \text{ (Secondary reaction)}$$

1. *Which ions are present in dilute sodium hydroxide solution?*
2. *Which of these ions is not discharged when sodium hydroxide solution is electrolysed?*
3. *When copper(II) sulphate is electrolysed with copper electrodes, no gas is produced at the positive electrode. The electrode does, however, reduce in size. Suggest how electrons are produced at this electrode.*
4. *Which ions are present in a concentrated solution of potassium iodide, KI, in water?*
5. *What are the products of the electrolysis of a concentrated solution of potassium iodide?*
6. *Suggest how the result in (5) would have been different if a very dilute solution of potassium iodide*

1. Na^+, H^+, OH^-; 2. Na^+; 3. $Cu \rightarrow Cu^{2+} + 2e^-$; 4. K^+, H^+, I^-, OH^-; 5. Iodine (at positive electrode) and hydrogen (at the negative electrode); 6. Oxygen may be collected instead of iodine.

13.3 Uses of electrolysis

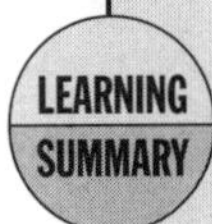

After studying this section you should be able to:

- ***explain the extraction of aluminium and the purification of copper by electrolysis***
- ***understand the difference in products obtained on using different electrolytes and electrodes.***

Electrolysis is used widely in industry.

Some uses include:

- **electroplating of metals**
- **electrolysis of brine**
- **purification of copper**
- **extraction of reactive metals such as sodium and aluminium.**

Electroplating of metals

This can be done for decorative purposes or to prevent corrosion.

Electrolysis can be used to put a thin coating of a metal on the surface of another metal.

If a piece of copper is to be plated with a layer of nickel, the copper must first be thoroughly cleaned and dried. **Copper** is made the **cathode** and a piece of

nickel is used as the **anode**. They are both dipped into **nickel(II) sulphate solution** (the **electrolyte**).

Nickel is deposited on the copper cathode.

$Ni^{2+} + 2e^- \rightarrow Ni$

The nickel anode goes into solution as nickel ions.

$Ni \rightarrow Ni^{2+} + 2e^-$

Electrolysis of brine

Electrolysis of brine (sodium chloride solution) produces sodium hydroxide, hydrogen and chlorine.

Fig. 13.8 shows a diaphragm cell into which concentrated brine is pumped.

In the cell, reactions occur at the electrodes.

At the titanium anode

$2Cl^-(aq) \rightarrow Cl_2(g) + 2e^-$

At the steel cathode

$2H^+(aq) + 2e^- \rightarrow H_2(g)$

If sodium hydroxide and chlorine react together sodium chlorate(I) is produced. This is used as household bleach.

The diaphragm keeps the products apart. The solution leaving the cell consist of 12% sodium hydroxide and 15% sodium chloride. Sodium hydroxide can be extracted from this solution.

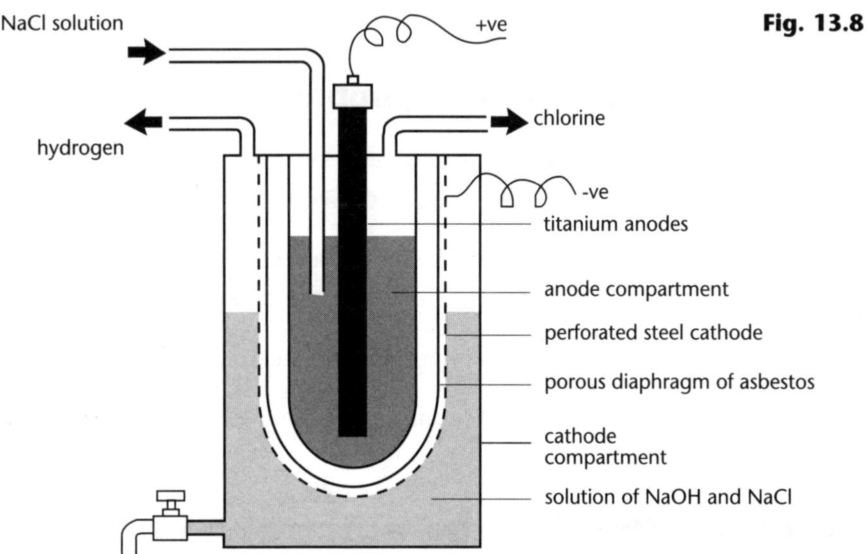

Fig. 13.8

PROGRESS CHECK

A brooch made of a copper alloy is going to be silver plated.

1. *What is the name of the process that can be used to do this?*
2. *Why is this better than just dipping it in molten silver?*
3. *Is the brooch made the anode or cathode in the process?*
4. *Suggest a metal for the other electrode.*
5. *Suggest a suitable electrolyte.*
6. *Write an equation for each electrode process.*

1. Electroplating; 2. Gives a thinner layer or a more even layer; 3. Cathode; 4. Silver;
5. A soluble silver salt, e.g. silver nitrate; 6. $Ag^+ + e^- \rightarrow Ag$; $Ag \rightarrow Ag^+ + e^-$.

Purifying metals by electrolysis

KEY POINT **Pure copper is a better electrical conductor than impure copper. It is an economic advantage to purify copper to a high purity.**

Copper is purified by **electrolysis** using the cell shown in **Fig. 13.9.**

A pure copper rod (called the **cathode** because it is connected to the negative terminal of the battery) and an impure copper rod (called the **anode**) are used. They are dipped into copper (II) sulphate solution (**electrolyte**).

KEY POINT **During the electrolysis, copper from the anode goes into solution as copper ions and copper ions from the solution are deposited on the cathode.**

These changes can be summarised by:

Anode $Cu \rightarrow Cu^{2+} + 2e^-$ (oxidised)

Cathode $Cu^{2+} + 2e^- \rightarrow Cu$ (reduced)

The impurities collect in the anode mud. As copper ores become rare and expensive, much of the new copper needed is obtained by **recycling** old copper wires and pipes.

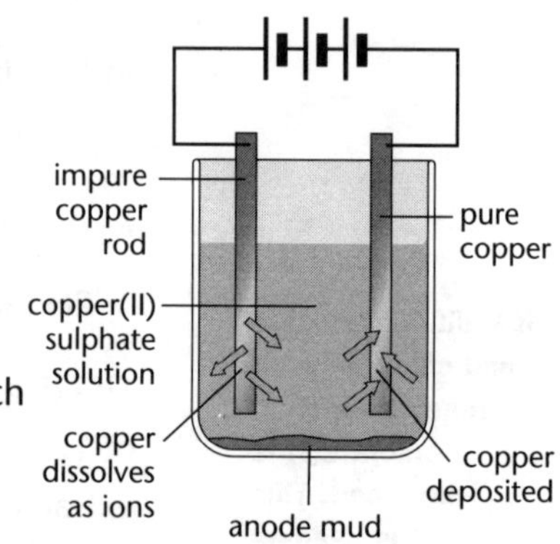

Fig. 13.9 Purification of copper

Uses of Copper

- Copper can easily be drawn into wires and being a good conductor, it is used in electrical wires.
- Copper being a good conductor of heat is used to make cooking utensils, as it facilitates uniform and fast heating in all parts of the utensil.
- Used in geysers and steam boilers.
- Used to make alloys, e.g. bronze and brass.

Extracting metals by electrolysis

This process of manufacturing aluminium is called Hall-Héroult process.

KEY POINT **Aluminium is extracted from purified aluminium oxide by electrolysis.**

Aluminium has a very high melting point and is not readily soluble in water. It does dissolve in molten cryolite(Na_3AlF_6). A **solution of aluminium oxide ($Al2O_3$)in molten cryolite** is a suitable electrolyte. The cell is shown in **Fig. 13.10**.

It takes the same amount of electricity to produce a tonne of aluminium as it does for all of the houses in a small town to use electricity for 1 hour.

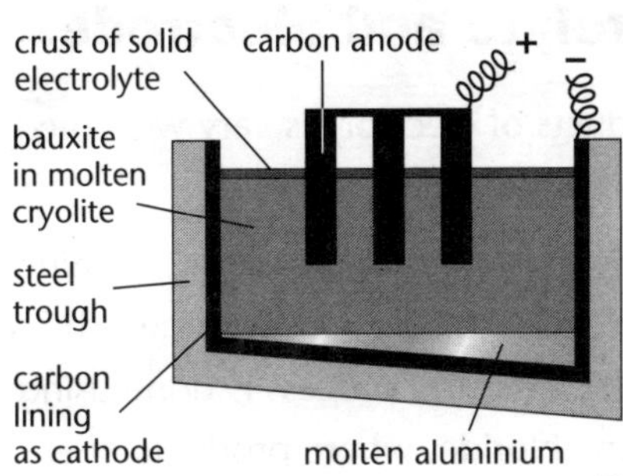

Fig. 13.10 Extraction of aluminium

In Anglesey in Wales an aluminium smelter uses electricity from the National Grid. A contract is negotiated to ensure electricity at an economic price.

The electrodes are made of carbon.

The reactions taking place at the electrodes are:

cathode $Al^{3+} + 3e^- \rightarrow Al$ (reduced)

anode $2O^{2-} \rightarrow O_2 + 4e^-$ (oxidised)

overall reaction:

$4Al^{3+} + 6O^{2-} \rightarrow 4Al + 3O_2$

As this process requires a large amount of electricity, an inexpensive source, e.g. hydroelectric power, is an advantage.

At the working temperature of the cell, the **oxygen** reacts with the **carbon** of the anode to produce **carbon dioxide**. The anode has, therefore, to be replaced frequently.

Uses of Aluminium

- Aluminium is used in the manufacture of food containers because of its resistance to corrosion.
- It can easily be drawn into wires and being a good conductor, is widely used in making electrical wires.
- Due to its high strength and low density, it is used in the manufacture of engines and aircraft. In particular, its alloy **duralumin** is widely used to manufacture aeroplane bodies.
- It is used in electrical cables because of low weight, chemical inertness and good conductivity.

PROGRESS CHECK

1. *Name a common ore for (i) iron (ii) aluminium?*
2. *What method is used to extract a metal at the top of the reactivity series?*
3. *Name a metal extracted using carbon monoxide.*
4. *What is tapped off from a blast furnace used for iron extraction, in addition to molten iron?*
5. *At which electrode is aluminium produced during its extraction?*
6. *Write an ionic equation for formation of aluminium from aluminium ions.*
7. *Why is cryolite used in aluminium extraction?*
8. *Write down the name of a metal purified by electrolysis.*
9. *Which metal is extracted by froth flotation process?*

1. (i) Haematite (or magnetite or iron pyrites) (ii) bauxite; 2. Electrolysis; 3. Iron (or zinc);
4. Slag (or calcium silicate); 5. Negative electrode (or cathode); 6. $Al^{3+} + 3e^- \rightarrow Al$;
7. As a solvent for aluminium oxide (or aluminium oxide has a very high melting point);
8. Copper; 9. Zinc.

Dependence of products of electrolysis on electrolyte and electrodes

The products of electrolysis vary with the type of electrolyte and the type of electrodes used.

As already illustrated in table 13.2, if solution of copper sulphete ($CuSO_4$) is electrolysed using carbon electrodes, the results obtained are different from the case when the same process is done using copper electrodes. Similar is the case of sodium chloride, where products of electrolysis vary with concentration of sodium chloride as well as type of electrode used.

Use of copper and aluminium in electric cables

The two most widely used materials in the manufacture of electrical cables are copper and aluminium.

Aluminium is used in the manufacture of electrical cables because of it's low density (thus making the cables light in weight), chemical inertness (leading to longer life) and good electrical conductivity (reduced losses in transmission of electricity.)

Copper is used in electrical cables as it is a very good conductor of electricity in purified form. However, as it easily reacts with air and moisture copper cables are usually covered with non-conducting materials, e.g. plastics. Insulators such as plastics and ceramics are used in electrical poles, switches etc., to segregate electrical wires from each other (and hence avoid short circuiting) and to prevent leakage of current in the electrical poles. This helps in reducing electricity losses besides minimising the risk of electric shocks to users.

Sample IGCSE question

1. Electrolysis of molten calcium bromide, $CaBr_2$, produces calcium and bromine.

(a) Which product is formed at the cathode and which at the anode? **[2]**

Calcium is produced at the negative electrode (cathode) ✓
and bromine at the positive electrode (anode) ✓

Remember that metals are formed at the negative electrode.

(b) Write equations for the ions being discharged at each electrode. **[2]**

Cathode $Ca^{2+} + 2e^- \rightarrow Ca$ ✓
Anode $2Br^- \rightarrow Br_2$ ✓ $+ 2e^-$

Check that each equation is balanced.

(c) Why the electricity does not pass until the calcium bromide is melted? **[2]**

In the solid ions are present ✓ But they are held in fixed positions unable to move ✓.

A frequent error here is to state that ions are only formed when calcium bromide melts.

Exam practice questions

1. The table gives the products of the electrolysis of some electrolytes with carbon electrodes.

Electrolyte	**Ion discharged at positive electrode**	**Product at positive electrode**	**Ion discharged at negative electrode**	**Product at negative electrode**
Molten sodium chloride	*Cl^-*	*chlorine*	*Na^+*	*sodium*
Sodium chloride solution (conc)	Cl^-	chlorine	Na^+	sodium
Sodium sulphate solution	SO_4^{2-}	sulphur	Na^+	sodium
Copper(II) sulphate solution	SO_4^{2-}	sulphur	Cu^{2+}	copper

(a) Complete the table. **[6]**

(b) How would the products be different if copper(II) sulphate solution was electrolysed with copper electrodes? **[3]**

2. Aqueous copper(II) sulphate solution can be electrolysed using carbon electrodes. The ions present in the solution are as follows.

Cu^{2+} (aq), SO_4^{2-} (aq), H^+ (aq), OH^- (aq)

(i) Write an ionic equation for the reaction at the negative electrode (cathode). **[1]**

(ii) A colourless gas was given off at the positive electrode (anode) and the solution changes from blue to colourless.

Explain these observations. **[2]**

Chapter 14

Experimental techniques

The following topics are covered in this section:

• Measurement • Criteria of purity • Methods of purification

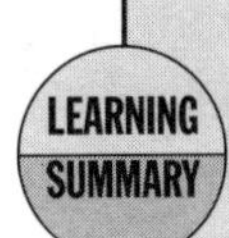

After studying the chapter, you should be able to:

- ***describe various measurement tools for time, temperature, mass and volume***
- ***explain the criteria of purity***
- ***describe various methods of purification.***

Measurement

In order to perform various experiments, we must be familiar with various apparatus used in measuring a given entity.

- Time is usually measured using a clock or a stopwatch.
- Temperature is measured using a thermometer.
- Mass is measured using a weighing balance, which may be electronic (digital) or manual.
- Volume is measured using either burettes, pipettes or measuring cylinders, all of which are graduated so that the exact volume of liquid is known.

Criteria of purity

Paper Chromatography

Chromatography is used for separation of two or more soluble solids from a mixture. This method is mostly used for identifying the constituents of a solution.

There are several types of chromatography, the simplest of them is **plain paper** chromatography.

To determine constituent dyes of say black ink, a sample of black ink is carefully spotted onto chromatography paper on a 'start line'.

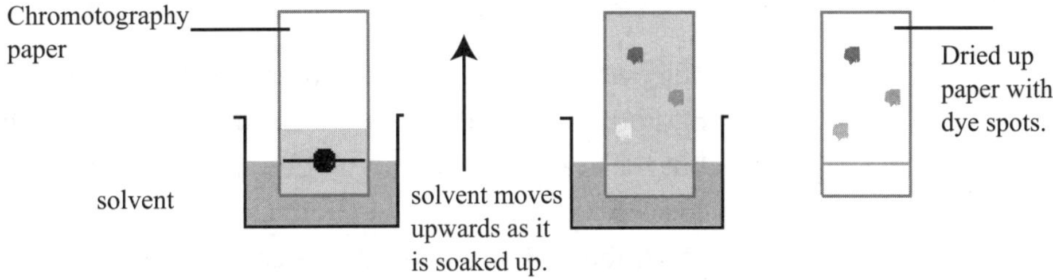

Fig. 14.1 Plain paper chromatogrphy

The paper is carefully dipped into a solvent, which is absorbed into the paper and rises up.

Due to different solubilities and different degrees of absorption by the chromatography paper, some colours move up the paper more than others effecting the separation of the different coloured molecules.

The distance a substance moves, compared to the distance the solvent front moves, is useful for analysis and identification. This method is used to analyse chromatograms by determining R_f values of the constituents (R_f value is the ratio of distance travelled by the solute to the distance travelled by the solvent).

It is also possible to analyse even colourless mixtures by this technique.This is done by spraying the chromatogram with a chemical called **locating agent**. The agent reacts with the colourlees substance to form a coloured substance and the spots become visible.

E.g. Proteins can be broken down into **amino acids** and **coloured purple** by a chemical reagent called ninhydrin.

Purity

- If only one substance is present in a material, then it is called **pure**. The pure material can either be an element or a compound but not a mixture.
- A simple physical test for purity is to measure the boiling point or melting point. A pure liquid or solid can be identified by its characteristic melting or boiling point.
- A pure solid has a sharp melting point.
- In a pure liquid, the temperature remains steady at its boiling point.
- On a chromatogram, a pure substance produces only one well-defined spot.

KEY POINT **Under given pressure, every pure substance melts and boils at a fixed temperature.**

- A pure substance becomes **impure** if one or more other substances are physically mixed into it.
- An impure liquid will boil at a comparatively higher or lower temperature than the pure one.
- An impure solid usually melts below its expected melting point.

The percentage purity of a compound is important, particularly in manufacture of drugs and vaccines. Impurities not only decrease the curative power of drugs but they can also be harmful. Laws have been enacted in most of the countries to keep a check over the impurities in drugs, vaccines, food items, cosmetics etc. Impurities also make things costly as the actual quantity of the substance purchased is less.

Methods of purification

KEY POINT **Purification is the process of separating the desired material from unwanted materials or impurities.**

Materials can be purified by various separation techniques, some of which are:

- **Decanting** - To separate an insoluble solid, settled at the bottom of a container, from a liquid or solvent.
- **Filtration** -To separate a solid from a liquid.
- **Evaporation** - To separate a dissolved solid from a liquid having a higher boiling point.
- **Distillation** - To separate a pure liquid from dissolved solid impurities, which have a higher boiling point.
- **Seperating funnel** - To separate mixtures of immiscible liquids.
- **Fractional distillation** - To separate mixtures of miscible liquids.
- **Centrifugation** - To separate suspended solids from a solid-liquid mixture.
- **Crystallisation** - To separate a dissolved solid from a liquid.
- **Chromatography** - To separate two or more soluble solids from a mixture.
- **Magnetic seperation** - To separate a magnetic solid, like iron, from a mixture.
- **Use of suitable solvent** - To separate a solid from a mixture of solids.

For methods of collecting gases see chapter 11.

Filtration

- It is one of the most simple and common methods of seperating insoluble solids from a solid-liquid mixture or a suspended solid from or a solid-gas mixture.
- The mixture is passed through a sieve or filter paper – the liquid along with the finer particles passes through the filter paper while the coarser particles are retained on it.

Distillation

The process involves seperation of a liquid from dissolved impurities by heating the mixture in a container and guiding the resulting liquid vapours through a cooling chamber(**condenser**). After passing through the condenser, the vapours are reconverted to liquid and collected in a separate container, whereas the impurities are left behind.

Fig. 14.2 shows an apparatus commonly used for distillation.

This process is used to purify water because the dissolved solids have a much higher boiling point and will not evaporate with the steam.

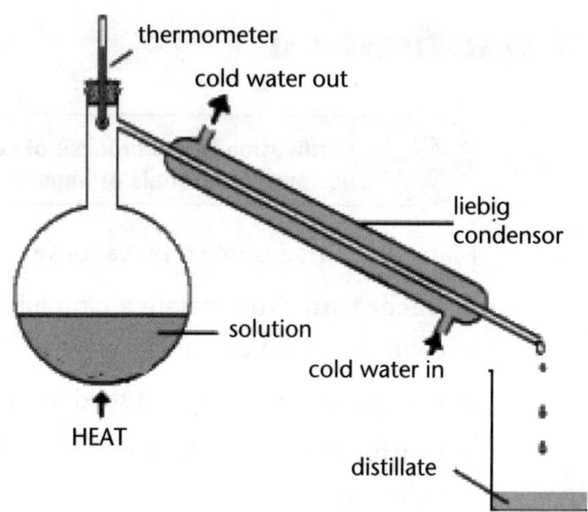

Fig. 14.2 Distillation process.

This method cannot be used to separate a mixture of liquids especially, if the boiling points are relatively close.

Fractional distillation

This method is used to separate two or more miscible liquids, having different boiling points, from a solution.

This can be used to separate liquor (alcohol) from a fermented solution.

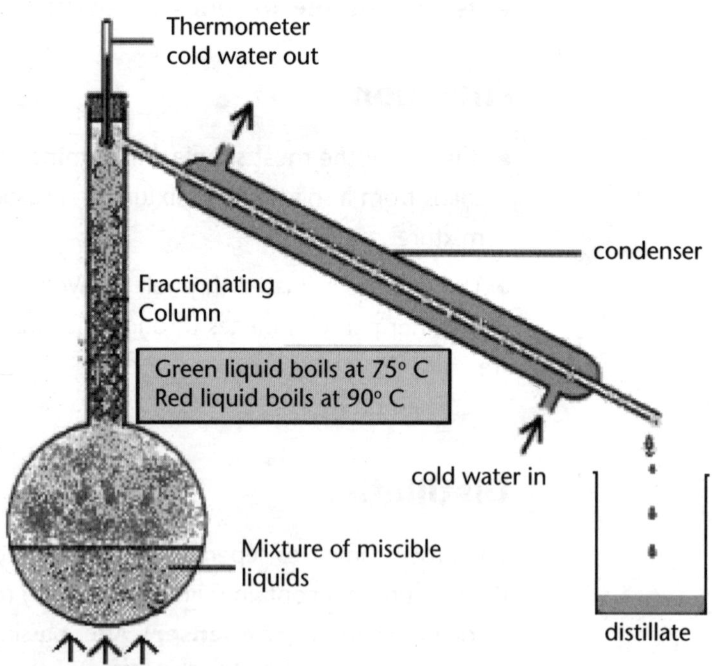

Fig. 14.3 Fractional distillation process.

1. The liquid or solution mixture is **boiled to vaporise** the most volatile component in the mixture.
2. The vapour passes up through a **fractionating column**, where the separation takes place.

3. The vapour is cooled by cold water in the condenser, to **condense** it back to
a liquid (the distillate) which is collected.

It is used on a large scale **to separate the components of crude oil** because different hydrocarbons have different boiling and melting points (See section 3.1- Refining crude oil)

Crystallisation

This is one of the most common methods of obtaining salt from sea water.

The process involves seperation of a dissolved solid from a solution by gradually drying up the solution to such an extent that a saturated solution is left behind. The concentration of this solution is so high that the solid begins to **crystallise** and it is seperated from the solution.

Use of suitable solvent

To separate a soluble solid from a solid-solid mixture, the soluble solid is dissolved in its solvent and seperated.

E.g. Grease stains are removed from cloth by dissolving the grease in a suitable solvent like petrol. The cloth being insoluble in petrol remains intact.

Sample IGCSE questions

1. Fill in the blanks for common measuring instruments, given below:

(a) The device given below is used for measuring __________ **[2]**

Time ✓

Name: ______________

Units: ______________

Stopwatch, seconds ✓

(b) The device given below is used for measuring ________________ **[2]**

Temperature ✓

Name: ________________

Units: __________

Thermometer, degrees ✓

(c) The device given below is used for measuring ___________ **[3]**

Weight ✓

(i) Name: ____________________________ **(ii)** Name: ____________________________

Units: ____________________________ Units: ____________

Physical balance, mg ✓ Electronic balance, mg ✓

(d) The devices given below are used for measuring __________ **[1]**

Volume ✓

Sample IGCSE questions

(i) Units for the following five apparatus: ______________. **[5]**

ml ✓

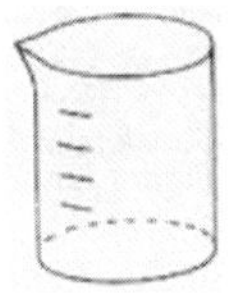

(ii) Name: __________

(approximate measurement)

beaker ✓

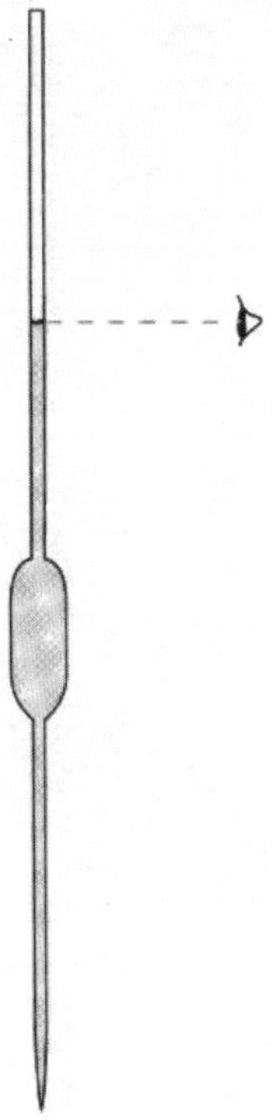

menisars

(iii) Name: ____________

(very accurate)

pipette ✓

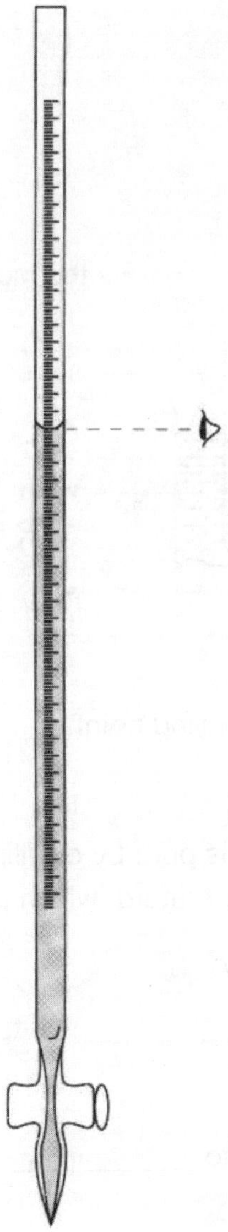

(iv) Name: ____________

(very accurate)

burette ✓

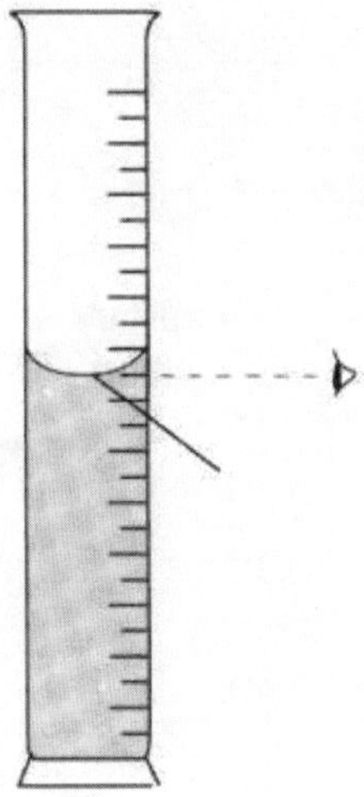

(v) Name: ____________

(more accurate)

measuring cylinder ✓

Sample IGCSE questions

2. Fill in the blanks **[13]**

(a) *Melting Point Determination*

(i) Impurities ____________ the melting point.

decrease ✓

(ii) The greater the percentage of impurity, the ______________ the melting point.

lower ✓

(iii) We can find out if a substance is pure, by measuring its melting point and then compare the measured value to the true melting point (recorded in books). This method is usually applied to _____________.

solids ✓

(b) *Boiling Point Determination*

(i) Impurities ____________ the boiling point.

increase ✓

(ii) We can find out if a substance is pure by distilling it. If the substance is pure, all of it distils at a _________ temperature, which is the ______________________.

constant ✓boiling point ✓

(iii) Otherwise, it will distil over a ___________ of temperatures.

range ✓

(iv) This method is usually applied to ___________.

liquids ✓

Sample IGCSE questions

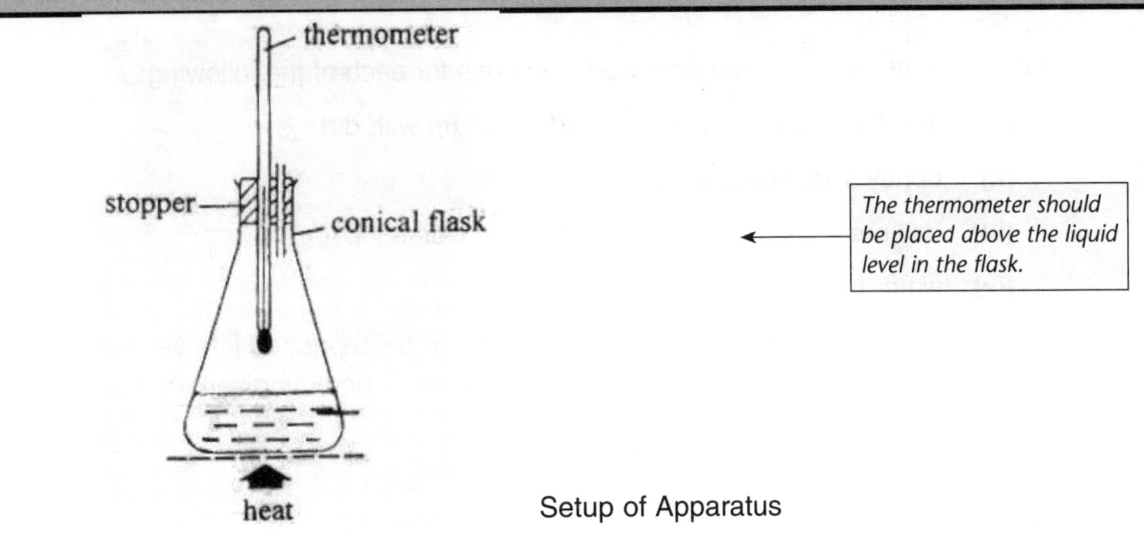
thermometer
stopper
conical flask
heat
The thermometer should be placed above the liquid level in the flask.

Setup of Apparatus

(c) *Chromatography*

(i) This method is used for complicated mixtures and those substances that do not melt or cannot be distilled. It is a very sensitive test and is able to detect very ____________________ of impurity.

small quantities ✓

(ii) As shown by the diagram below, a pure substance gives ______ spot on a chromatogram.

single ✓

(iii) A mixture gives ____________________ on a chromatogram.

multiple spots ✓

Chromatography can also be used to separate and identify colourless substances.

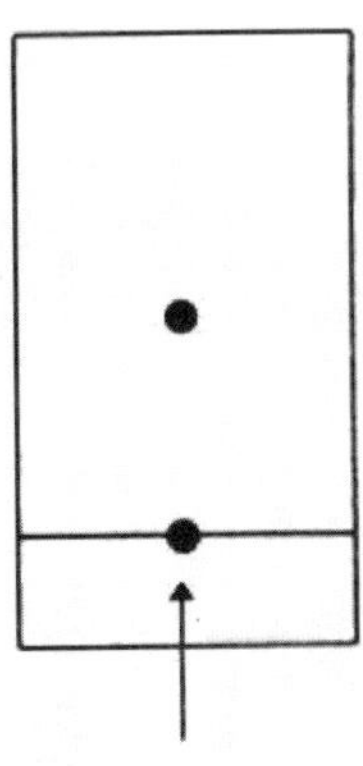

original sample
Chromatogram of a pure substance

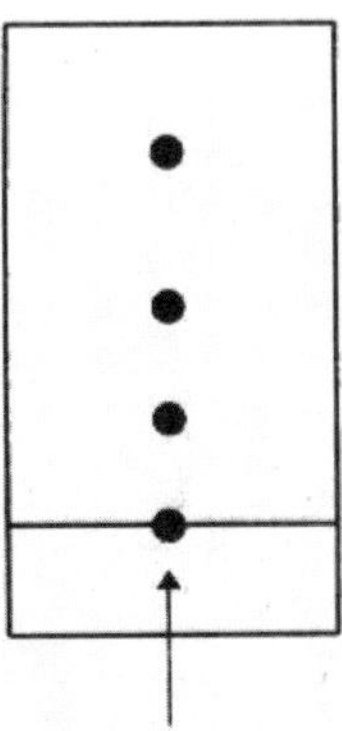

original sample
Chromatogram of a mixture

(iv) The chromatogram is sprayed with a __________________________ to show where the substances are on the paper. The locating agent is a chemical, that reacts with the substances to produce a ________________ product.

locating agent ✓ coloured ✓

Exam practice questions

1. (a) What method of separation would you use for each of the following: **[4]**

(i) iron filings spilt on the floor and swept up with dirt

(ii) honey from honeycomb

(iii) ball bearings from a vat of oil in a car workshop

(iv) water from wet clothes?

(b) Milk is often described as being 'pure' but can be separated into various components, using a centrifuge. How, then, can it be advertised as pure milk? **[2]**

Exam practice answers

Chapter 1 The particulate nature of matter

1 2 Answers should include something about diffusion of particles. **[2] + [2]**

3 Liquids get converted to solids as the particles loose energy because of cooling. The attractive forces binding the particles together are strengthened and the particles are not able to move around freely. This results in a solid shape. **[2]**

Chapter 2 Atoms, elements and compounds

1 (a) (i)

One mark for the pair of electrons between the two atoms.
One mark for other electrons correctly shown. **[2]**

(ii) Covalent bond **[1]**

(b) Covalent bond breaks
Forms two ions
H^+ and Cl^-
Electricity is transferred by ions.
Any three points. **[3]**

2 Substance A has a giant structure of ions.
Giant structure because of high melting point and conducts electricity when molten but not when solid.
Substance B has a metallic structure.
Conducts electricity when solid.
Substance C has a molecular structure.
Low melting and boiling point.
Substance D has a giant structure of atoms.
High melting point and does not conduct electricity even in molten state **[8]**

3 (a) 2, 8, 1 **[1]**
(b) 2, 8, 8, 2 **[1]**

4 Elements A and B are isotopes. **[1]**

5

Element	Proton Number (At. No.)	Nucleon Number (At. Mass)	Valency	Number of Protons	Number of Neutrons	Number of Electrons
O	8	16	-2	**8**	8	**10**
N	7	14	**-3**	**7**	7	10
F	9	18	-1	**9**	**9**	10
K	19	39	+1	**19**	**20**	**18**

One mark for each answer (shown in bold).

6 Oxygen has 8 electrons which means that it requires 2 more electrons to complete its octet. **[1]** However, if it has to lose electrons, it will have to lose 6 which is difficult. **[1]**

7 (a) Any use that depends on malleability or ductility like jewellery, pipes, wires, sheets, roofing, ornaments. **[1]**
(b) Electrical wires or cooking utensils or electrodes **[1]**
(c) Making alloys **[1]**

8 (a) -3 **[1]** (b) +1 **[1]** (c) zero **[1]** (d) 10 **[1]**

Chapter 3 Organic chemistry

1 (a) (i) carbon, hydrogen and oxygen – 3 correct **[2]**; 2 correct **[1]**
(ii) Fermentation **[1]**
(iii) Yeast or enzymes or zymase **[1]**
(b) Fractional distillation of crude oil **[1]**
Cracking (or thermal decomposition) **[1]**
Of high boiling point fractions **[1]**
Using high temperature and catalyst **[1]**
(c) Making poly(ethene) **[1]**
(d) ethene + water → ethanol **[1]**
$C_2H_4 + H_2O \rightarrow C_2H_5OH$ **[2]**
(e) (i) Oxygen is added. **[1]**
(ii) Cold air enters the apparatus and warm air leaves. **[1]**

2 (a) $C_{14}H_{18}O_5N_2$ **[1]**
(b) Not a carbohydrate. Does not fit formula for carbohydrate. **[1]**
(c) Does not have the high energy content of a carbohydrate. **[1]**

3 —O—☐—O—☐—O—☐—O— **[1]**
chain **[1]**

4 (a) (i) in which something dissolves **[1]**
(ii) $CH_3COOC_2H_5$ **[1]**
(iii) steam or water **[1]**
heat or catalyst **[1]**
oxidised **[1]** by air or dichromate or manganate(VII) **[1]**
(iv) ethanoic acid and methanol **[1]**
(b) (i) CH_2OH
CHOH
CH_2OH **[1]**
(ii) Soap or detergent **[1]**

Chapter 4 Carbonates

1 Limestone, quicklime or slaked lime (any two chemicals). **[2]**

2 Carbon dioxide **[1]**

3

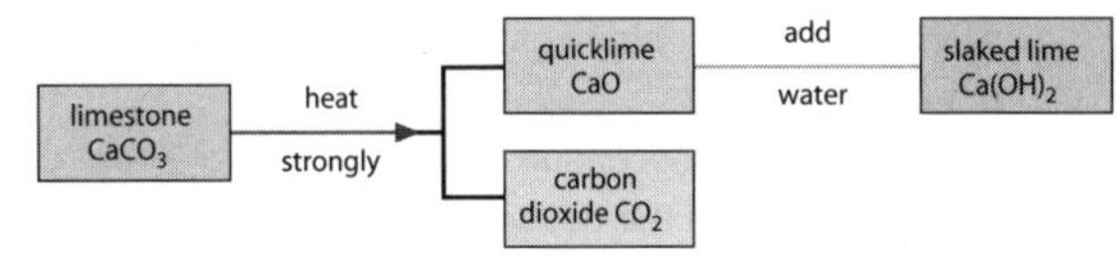

One mark for each chemical. **[3]**

Chapter 5 Stoichiometry

1 (a) (i) Pipette **[1]**
(ii) Burette **[1]**
(b) An indicator **[1]**
e.g. phenolphthalein, which changes from colourless to pink. **[1]**
(c) (i) $35/1000 \times 1.2 = 0.042$ **[1]**
(ii) $0.5 \times 0.042 = 0.021$ **[1]**
(iii) $1000/25 \times 0.021 = 0.84$ **[1]**

2 (a) $\frac{20 \times 0.1}{1000}$ **[1]** = 0.002 **[1]**
(b) 0.002 **[1]**
Note: The equation shows that 1 mole of ethanoic acid reacts with 1 mole of sodium hydroxide.
(c) 0.04 moles **[1]**
(d) 60 g **[1]**
(e) 0.04×60 **[1]** = 2.40 g/dm^3 **[1]**

3 (a) Avogadro's Number of particles or formula mass in grams **or** 6×10^{23} particles like atoms, ions and molecules **or** as many particles as there are carbon atoms in 12.00 g of ^{12}C (ANY one). **[1]**

(b) (i) moles of Mg = 5/24 = 0.208
moles of CH_3COOH = 20/60 = 0.333
magnesium is in excess **[1]**

(ii) moles of H_2 = 0.333/2 = .166 **[1]**

(iii) Volume of hydrogen = 0.166 × 24 = 3.984 dm^3 **[2]**

(c) (i) moles of NaOH = 30/1000 × 0.5 = 0.015 **[1]**

(ii) moles of acid = 0.015/2 = 0.0075 **[1]**

(iii) concentration of acid = 0.0075 × 1000/25 **[1]**
= 0.3 mol/dm^3 **[1]**

Chapter 6 Air and water

1 (a) (i) different boiling points **[1]**

(ii) by reaction between methane and steam with nickel catalyst **[1]**
$CH_4 + H_2O \rightleftharpoons 3H_2 + CO$ **[1]**

(b) (i) Volume decreases for forward reaction or fewer moles of gas on products side **[1]**
Favoured by increase in pressure **[1]**

(ii) increase **[1]**
exothermic reaction favoured by lower temperature **[1]**

(iii) 300 – 600 degrees C
1:3 volume ratio
iron (catalyst)
150 to 300 atm (any two) **[2]**

(c) (i) proton or
hydrogen ion **[1]**

(ii) correct equation either molecular or ionic **[2]**

Chapter 7 The periodic table

1 (a) Group 2 metals are less reactive than corresponding group 1 metals. **[1]**
Group 2 metals have 2 electrons in outer shell and group 1 have 1 electron. **[1]**
More difficult to lose 2 electrons than 1. **[1]**

(b) Reactivity increases down group. **[1]**
Atoms are larger down the group. **[1]**
Easier to lose electrons further from the nucleus. **[1]**

2 (a) (i) Reactivity increases down the group. **[1]**

(ii) Atoms of all elements of group 2 have two electrons in the outer shell **[1]**
As the group is descended these two electrons are farther away from the nucleus **[1]**
Force of attraction between nucleus and electrons becomes weaker. **[1]**
Electrons are more easily lost. **[1]**

(b) (i) $CaCl_2$ **[1]**

(ii) Add dilute hydrochloric acid **[1]**
To a measured amount of calcium hydroxide with indicator **[1]**
Until indicator changes colour. **[1]**
Repeat without indicator. **[1]**
Evaporate until small volume of solution remains. **[1]**
Leave to cool and crystallise. **[1]**
Any five points

Chapter 8 Chemical reactions

1

New condition	Change, if any	Explanation
Use 5 g of powdered zinc	Faster **[1]**	Larger surface area **[1]**
Use 40 °C	Faster **[1]**	Higher temperature, particles move faster – more collisions **[1]**]
Use 100 cm^3 of hydrochloric acid (50 g/dm^3)	Slower **[1]**	Lower concentration – fewer collisions between acid particles and zinc **[1]**
Use 100 cm^3 of ethanoic acid (100 g/dm^3)	Slower **[1]**	Ethanoic acid is a weak acid – only partially ionised **[1]**

2 (a) 80 **[1]**
80 or 160 **[1]**
2 **[1]**

(b) particles have more energy or moving faster **[1]** collide more frequently **[1]**

(c) greater surface area **[1]**

(d) flour mills or coal mines or metal powders **[1]**

3 (a) (i) collect or measure volume or mass of oxygen **[1]**
time **[1]**

(ii) measure rate in different intensities of light and comment **[1]**

(b) $6O_2$ **[2]**

Chapter 9 Sulphur

1 copper pyrite, zinc blende, lead sulphide **[Any 2]**
2 Froth floatation process **[1]**
3 Contact process **[1]**
4 Hypo (sodium thiosulphate) **[1]**
5 Sulphur dioxide **[1]**

Chapter 10 Chemical changes

1 (a) Reactions which absorb energy from the surroundings are called endothermic reactions **[1]** whereas reactions which give out energy to the surroundings are called exothermic reactions. **[1]**

(b) Endothermic reaction **[1]**

2 The main advantage of cells and batteries is that they can be constructed in very small sizes **[1]** and are therefore widely used to power portable electronic and electrical devices like torch, radio, watches etc. **[1]**

Chapter 11 Acids, bases and salts

1 This is a sample answer. There would be other ways of doing it.

(a) Add dilute hydrochloric acid and barium chloride to each solution **[1]**
Sodium sulphate and sulphuric acid produce white precipitates. **[1]**
Sodium chloride does not form a precipitate and can be identified. **[1]**
Add litmus to samples of sodium sulphate and sulphuric acid. **[1]**
Sulphuric acid turns litmus red **[1]**
Sodium sulphate turns litmus purple. **[1]**

(b) $Ba^{2+}(aq) + SO_4^{2-}(aq) \rightarrow BaSO_4(s)$ **[2]**

2 Use solution of soluble lead(II) salt **[1]**
e.g. lead(II) nitrate. **[1]**
Mix solutions **[1]**
Filter off lead(II) chromate **[1]**
Wash with distilled water **[1]**
1 mark for quality of written communication.
$Pb(NO_3)_2 + K_2CrO_4 \rightarrow PbCrO_4 + 2KNO_3$ **[3]**

3 (a) $NH_4Cl + NaOH \rightarrow NH_3 + NaCl + H_2O$ **[1]**
(b) (i) Ammonia dissolves in cold water **[1]**
(ii) Use damp red litmus **[1]**
Hold it near the mouth of the test tube **[1]**
Test tube full of ammonia when the litmus turns blue **[1]**

4 (a) Measure pH or add universal indicator or pH meter. **[1]**
Ammonia has lower pH if numericals values given are appropriate (that is, above 7) with ammonia having the lower value. **[1]**

OR

Measure conductivity **[1]**

Ammonia has poorer conductivity. **[1]**

(b) (i) Correct structural formula **[1]**

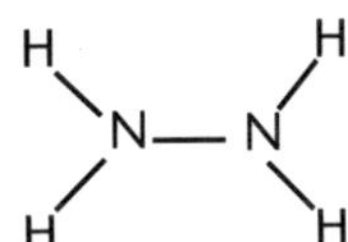

(ii) 8e⁻ around nitrogen **[1]**
2e⁻ around each hydrogen **[1]**

H H
xo xo
H $\overset{x}{o}$ N $\overset{x}{x}$ N H
xx xx

1 mark for dot and cross diagram.

Chapter 12 Metals

1 Name of iron ore e.g. haematite
Iron ore, limestone and coke added to the blast furnace.
(heated with hot air) **[1]**
Iron tapped off the bottom of the furnace (as pig iron contains unwanted impurities) [1]
Pig iron (and scrap iron) and limestone added
and heated until molten **[1]**
Oxygen (or air) blown onto the surface of the molten iron/Oxidises impurities/removed as slag [1]
Answers in () represent optional answers whereas rest must be present to secure the full mark. For the fourth mark, there is a choice of three answers (separated by slashes).

2 (i) heat or roast **[1]**
in air **[1]**
(ii) Correct balanced equation
$ZnO + C = Zn + CO$ **[2]**
(iii) BP of lead above 1400 °C – it remains; BP of zinc below 1400 °C – boils away **[2]**
OR lead does not boil and zinc boils **[2]**

3 (a) (i) Bodies of airoplane or overhead power cables **[1]**
(ii) lattice [1]
free delocalised mobile electrons **[1]**
layers/atoms/particles can slip **[1]**
(iii) atoms of different sizes **[1]**
prevents layers from moving **[1]**
(b) (i) one involving lead – change 2 **[1]**
because electrons are gained or oxidation number is reduced **[1]**
(ii) correct equation $Zn + 2Ag^+ \rightarrow 2Ag + Zn^{2+}$ **[2]**

Chapter 13 Electrochemistry and electrolysis

1 (a) Cl^-; chlorine, H^+; hydrogen; OH^-; oxygen; H^+; hydrogen; OH^-; oxygen; Cu^{2+}; Cu. Half mark for each correct answer – rounded up to whole mark. **[6]**
(b) At the anode (positive electrode) **[1]**
Copper goes into solution **[1]**
As copper(II) ions **[1]**

2 (i) $Cu^{2+} + 2e^- \rightarrow Cu$ **[1]**
(ii) gas is oxygen **[1]**
copper(II) sulphate changes to sulphuric acid **[1]**
or copper ions removed from solution

Chapter 14 Experimental techniques

1 (a) (i) magnetic **[1]**
(ii) sieving (slow) or centrifuging **[1]**
(iii) decanting or sieving (messy) **[1]**
(iv) sieving **[1]**
(b) Milk is a mixture of various substances including fat, water and protein. 'Pure' milk refers to the absence of contaminants, such as chemicals or microorganisms. **[2]**

INTERNATIONAL GENERAL CERTIFICATE OF SECONDARY EDUCATION

Model Test Paper

Chemistry (Paper 1 - Multiple Choice)

M.M. 40 **Total no. of Questions: 40**

Leave blank

1. Molten sodium chloride conducts electricity. The current is carried by

   a. molecules

   b. metal atoms

   c. ions

   d. non-metal atoms.

2. Potassium has an atomic number of 19. Which of the following represents the electronic configuration of the stable potassium ion K^+?

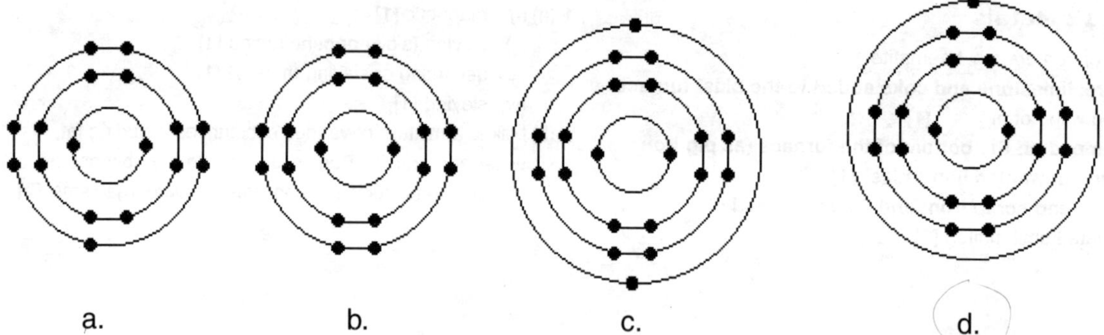

3. From the properties of the four substances described below, which is most likely to consist of a giant covalent structure?

   a. Melting point 1150°C, molten substance readily conducts electricity.

   b. Melting point 480°C, molten substance readily conducts electricity.

   c. Melting point 1500°C, molten substance is a poor electrical conductor.

   d. Melting point 20°C, molten substance doesn't conduct electricity.

4. Copper is a metallic element, so it is likely to be a

   a. poor conductor of electricity

   b. substance with a low melting point

   c. substance easily hammered into shape

   d. poor conductor of heat.

5. From the properties of the four substances described below, which is most likely to be an ionic compound?

   a. Melting point 150°C, molten substance doesn't conduct electricity.

   b. Melting point 800°C, molten substance readily conducts electricity.

   c. Melting point 1500°C, molten substance is a poor electrical conductor.

   d. Melting point 20°C, molten substance doesn't conduct electricity.

6. Sodium chloride is a typical ionic compound, formed by combining a metal with a non-metal. Sodium chloride will

   a. consist of NaCl molecules

   b. not conduct electricity when dissolved in water

   c. have a low melting point

   d. not conduct electricity, when solid.

7. A substance has a giant covalent structure and a high melting point. The substance could be

   a. aluminium

   b. copper

   c. polyethene

   d. carbon (diamond).

8. Sodium chloride has a high melting point because it has

   a. many ions bonded together

   b. strong double covalent bonds

   c. molecules tightly packed together

   d. a giant covalent three-dimensional structure.

Leave blank

Leave blank

9. Atoms of group 6 of the periodic table are most likely to form ions with

   a. a single negative charge by gaining one electron

   b. a double positive charge by losing two electrons

   c. a single positive charge by losing one electron

   d. a double negative charge by gaining two electrons.

10. Which of the following statements is true about the trend DOWN the group 7 (Halogens), with increase in atomic number?

    a. The reactivity increases.

    b. The atoms get smaller.

    c. The boiling points decrease.

    d. The melting points increase.

11. Which of the following is used in the manufacture of photographic film?

    a. Sodium hydroxide

    b. Silver chloride

    c. Chlorine

    d. Hydrogen

12. Which of the following does NOT represent a possible reaction?

    a. chlorine + sodium iodide → potassium chloride + iodine

    b. fluorine + sodium chloride → potassium fluoride + chlorine

    c. bromine + potassium iodide → potassium bromide + iodine

    d. iodine + potassium bromide → potassium iodide + bromine

13. Which of the following halogens (at room temperature and pressure) is a green gas?

    a. Iodine

    b. Bromine

    c. Chlorine

    d. Fluorine

Leave blank

14. The displacement reaction between chlorine and potassium bromide can be represented by the ionic-redox equation:

$$Cl_2(aq) + 2Br^-(aq) \rightarrow 2Cl^-(aq) + Br_2(aq)$$

Identify the correct statement.

a. Bromide is oxidised.

b. Chlorine is the reducing agent.

c. Bromide is the oxidising agent.

d. Chlorine is oxidised.

15. The equation for the burning of butane is shown below.

$$2C_4H_{10} + 13O_2 \rightarrow 8CO_2 + 10H_2O$$

How many moles of water are formed when one mole of butane burns completely?

a. 4

b. 5

c. 8

d. 10

16. Gases, unlike solids, may be easily compressed because

a. gas molecules are softer than those in solids

b. gas molecules are smaller than those in solids

c. gas molecules can move but those in solids cannot

d. gas molecules are far apart but those in solids are touching each other.

17. The table shows the densities for some of the group 1 metals.

metal	density (in g/cm^3)
a. lithium	0.53
b. sodium	0.97
c. potassium	0.86
d. rubidium	1.53

Which metal sinks in benzene (density of liquid = 0.88 g/cm^3) but floats in nitrobenzene (density of liquid = 1.2 g/cm^3)?

Leave blank

18. What happens when ice melts?

    a. Water molecules change to hydrogen and oxygen atoms.

    b. Water molecules change to water atoms.

    c. Irregularly arranged molecules change to regularly arranged molecules.

    d. Regularly arranged molecules change to irregularly arranged molecules.

19. Given the following cracking equation, what is the missing molecule?

$$C_9H_{20} \rightarrow C_7H_{16} + \_\_\_\_$$

    a. C_2H_4

    b. C_3H_8

    c. C_2H_6

    d. C_3H_6

20. Which of the following statements about crude oil is TRUE?

    a. It is a non-renewable energy resource.

    b. It is a compound.

    c. It is formed from the erosion of sedimentary rock.

    d. It is an infinite energy resource.

21. Which of the following separation methods would be used to remove yeast from fermented sugar solution?

    a. Chromatography

    b. Crystallisation

    c. Fractional distillation

    d. Filtration

22. Which of the following chemical reagents and reaction conditions are used for 'breakdown of starch into glucose'?

    a. Hydrogen gas and nickel catalyst

    b. Acidified potassium dichromate(VI)

    c. Steam and phosphoric(V) acid catalyst

    d. Heating with hydrochloric acid

Leave blank

23. Which noble gas is used in filament bulbs, to prevent the filament from burning out too quickly?

    a. Helium

    b. Argon

    c. Krypton

    d. Neon

24. In 1962, a Canadian chemist *Bartlett*, managed to combine xenon with the highly reactive halogen gas fluorine to make xenon tetrafluoride. Which equation depicts the reaction correctly?

    a. $Xe_2 + 8F_2 \rightarrow 2XeF_4$

    b. $Xe + 2F_2 \rightarrow XeF_4$

    c. $2Xe + 4F \rightarrow 2XeF_4$

    d. $Xe_2 + 2F_2 \rightarrow XeF_4$

25. Sodium and potassium would be expected to be very similar chemically because they both:

    a. have the same formula for a chloride salt

    b. have the same number of electrons in their respective atoms

    c. have the same number of outer electrons

    d. react quickly with water.

26. Which statement is TRUE about the reaction of lithium and water, containing universal indicator?

    a. The products are lithium oxide and hydrogen.

    b. The indicator turns from neutral green (pH 7) to weakly alkaline dark green (pH 8).

    c. The reaction is fast and endothermic.

    d. The gas formed gives a pop with a lit splinter.

27. Which of the following is a typical property of a transition metal?

    a. Soft solid

    b. Low melting point

c. High boiling point

d. Low density

28. What is the charge on the ion, formed by an alkali metal?

a. –1

b. +2

c. -2

d. +1

29. Which pair of elements consists of non-metals which will displace less reactive metals from their oxide ores?

a. Sulphur and oxygen

b. Chlorine and argon

c. Carbon and hydrogen

d. Nitrogen and helium

30. Which metal is added to aluminium to make a stronger alloy?

a. Sodium

b. Chromium

c. Gold

d. Magnesium

31. Which of these metals can be used in the 'sacrificial corrosion' method for protecting steel structures from rusting?

a. Iron

b. Zinc

c. Tin

d. Silver

32. Which of the following removes the oxygen from the iron ore in a blast furnace?

a. Limestone

b. Slag

c. Carbon dioxide

d. Carbon monoxide

Leave blank

Leave blank

33. When hydrochloric acid is neutralised with sodium hydroxide, which two ions have the lowest concentrations in the resulting mixture?

    a. H^+ and Cl^-

    b. Na^+ and Cl^-

    c. Na^+ and OH^-

    d. H^+ and OH^-

34. Which reaction causes the high temperature in a blast furnace?

    a. carbon + oxygen → carbon dioxide

    b. calcium carbonate + silicon dioxide → calcium silicate + carbon dioxide

    c. carbon dioxide + carbon → carbon monoxide

    d. iron(III) oxide + carbon monoxide → iron + carbon dioxide

35. After the initial chemical extraction of copper from a copper ore, the copper is purified by:

    a. distillation

    b. electrolysis

    c. hydrolysis

    d. crystallisation.

36. Given are the following four observations of the reactions of four metals.

    i. Metals F and G react slowly with water.

    ii. Metal H displaces metal G from its chloride salt solution.

    iii. Metal E does not react with dilute acids.

    iv. Metal H does NOT displace metal F from its sulphate salt solution.

    What is their order of reactivity, starting from the most reactive metal to the least reactive one?

    a. F > H > G > E

    b. H > G > F > E

    c. H > E > F > G

    d. E > G > H > F

Leave blank

37. Rust protection by galvanising steel car bodies is done by coating the metal with:

    a. hard wearing paint

    b. tin

    c. copper

    d. zinc.

38. Which statement is TRUE, if copper and zinc metal electrodes are used in a simple battery cell made by dipping strips of zinc and copper into their salt solutions?

    a. $Cu(s) \rightarrow Cu^{2+}(aq) + 2e^-$ occurs on the copper plate

    b. The zinc electrode dissolves slowly.

    c. The zinc electrode plate will turn red-brown in colour.

    d. $Zn^{2+}(aq) + 2e^- \rightarrow Zn(s)$ occurs on the zinc plate.

39. This table shows the melting points and boiling points of four substances.

Substance	Melting point /°C	Boiling point /°C
a.	–203	–17
b.	–25	–50
c.	11	181
d.	463	972

    Which substance is liquid at 100 °C?

40. An element has the electronic structure 2,8,3. It may be deduced that the element

    a. is in group 2 of the periodic table

    b. is in group 8 of the periodic table

    c. is in group 3 of the periodic table

    d. has a relative atomic mass of 11.

INTERNATIONAL GENERAL CERTIFICATE OF SECONDARY EDUCATION

Model Test Paper

Chemistry (Paper 3 - Extended)

M.M. 80 **Total no. of Questions: 8**

Leave blank

1 Copper(II) oxide is added to dilute sulphuric acid in a test tube and the mixture warmed. When the test tube is left in a rack for a few minutes a black solid settles to the bottom, leaving a clear blue liquid above.

(a) (i) Write down the name of the clear blue liquid.

.... **[1]**

(ii) Write down the name of the black solid in the test tube.

.... **[1]**

(iii) Write a balanced equation for the reaction between copper(II) oxide and sulphuric acid.

.... **[2]**

(iv) What name is given to this type of reaction?

.... **[1]**

(b) Describe how you would make blue crystals from the mixture in the test tube. (One mark is for the correct sequencing of your answer.)

....

....

....

.... **[3+1]**

(c) The reaction between copper(II) oxide and sulphuric acid gives out energy in the form of heat.

(i) What name is given to a reaction which gives out heat?

.... **[1]**

(ii) Draw an energy level diagram to represent this reaction.

[3]

(Total 13 marks)

Leave blank

2 The apparatus shown in the diagram was used to study the combustion of a liquid hydrocarbon, octane, C_8H_{18}.

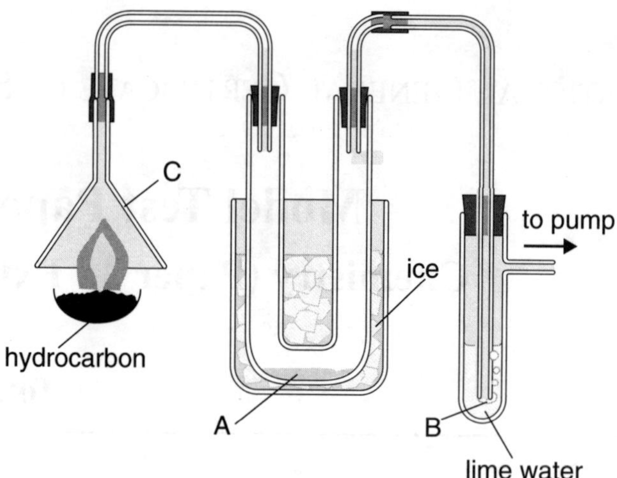

Gases from the burning hydrocarbon were drawn through the apparatus for several minutes.

(a) A clear, colourless liquid appeared at A.

(i) Name this liquid.

.... **[1]**

(ii) Describe a test to prove the identity of this liquid.

....

.... **[2]**

(b) (i) What would you see at B, as the experiment progresses?

....

.... **[2]**

(ii) What does this show about the gases produced by combustion of the hydrocarbon?

.... **[1]**

(c) (i) Write a balanced equation for the complete combustion of octane.

.... **[1]**

(ii) When the experiment was completed, a black deposit of carbon was noted at C.
Explain how this was formed.

....

.... **[2]**

(Total 9 marks)

3 Sarah studied the reaction between hydrochloric acid and sodium carbonate.

$$2HCl + Na_2CO_3 \rightarrow 2NaCl + CO_2 + H_2O$$

She made hydrochloric acid of different concentrations by mixing a more concentrated solution with water. She used tablets, each of which contained the same mass of sodium carbonate. She timed how long it took for a tablet to react completely in the same volume of each concentration of acid.
Her results are shown in the table.

volume of acid in cm^3	volume of water in cm^3	time for tablet to react in seconds
2	18	350
4	16	245
6	14	220
8	12	142
10	10	57

(a) Plot the volume of acid used against time on the grid below.
Draw the line of best fit for the points you have plotted. **[3]**

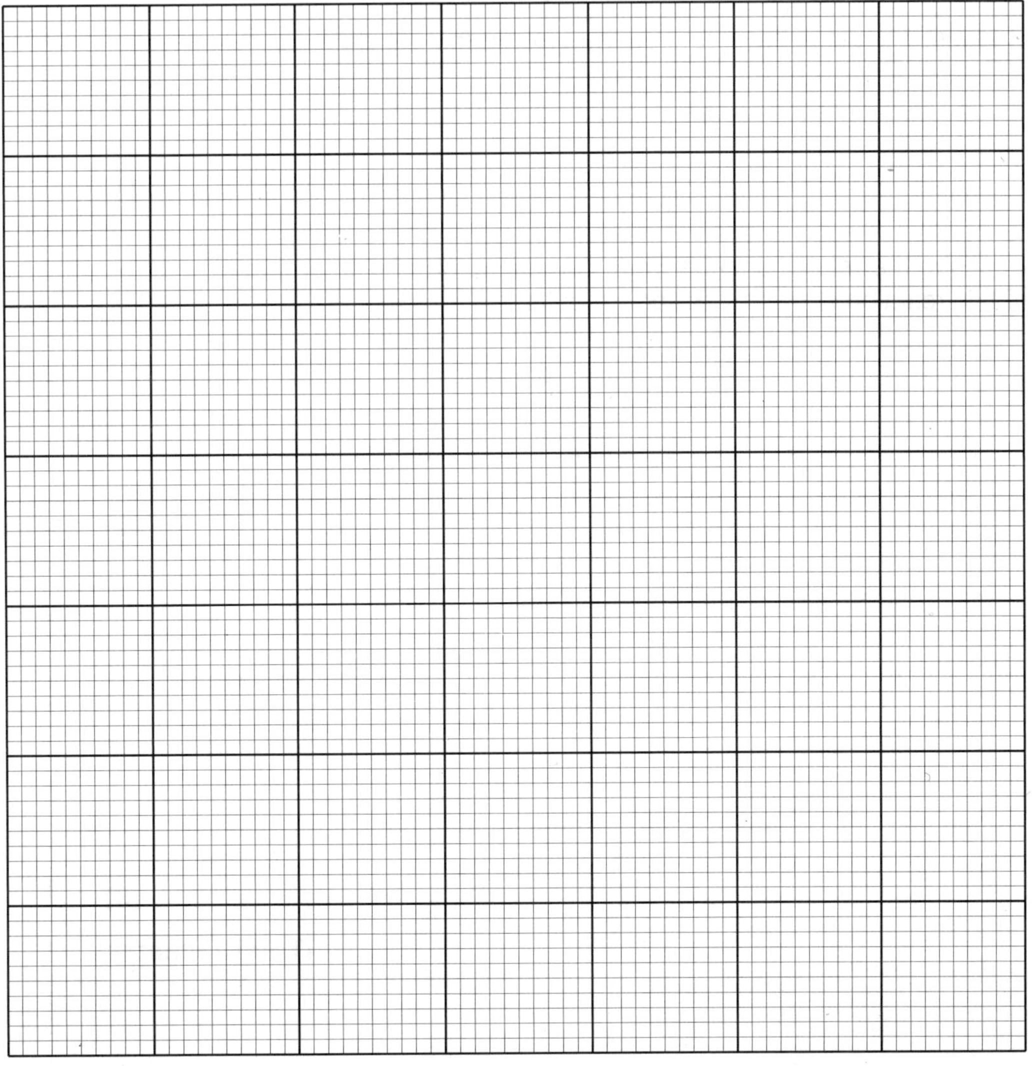

Leave blank

Leave blank

(b) Sarah was careful to ensure fair testing in her experiments.

(i) Explain how mixing each volume of acid with a different volume of water helped to ensure fair testing.

....

....

.... **[2]**

(ii) What other thing, not mentioned above, must Sarah have kept constant to ensure fair testing?

.... **[1]**

(c) One of Sarah's results is anomalous.

(i) What volume of acid was used for the anomalous result?

.... **[1]**

(ii) Suggest what may have caused the error in this result.

....

.... **[1]**

(d) **(i)** Describe the relationship between concentration of acid and rate of this reaction shown by Sarah's results.

....

....

.... **[2]**

(ii) Use your knowledge of particles to explain this relationship.

....

....

....

.... **[3]**

(Total 13 marks)

Leave blank

4. Ammonia is manufactured from nitrogen and hydrogen.

$$N_2(g) + 3H_2(g) \rightleftharpoons 2NH_3(g)$$

(a) What does the symbol $\rightleftharpoons$ show?

.... **[1]**

(b) Where are the raw materials nitrogen and hydrogen obtained from?

nitrogen is obtained from ....

hydrogen is obtained from .... **[2]**

(c) Suggest two reasons why the manufacture of ammonia is carried out at high pressure.

1 ....

2 .... **[2]**

(d) At low temperatures, a very high yield of ammonia can be obtained if the mixture is left long enough. Explain why the process is actually carried out at higher temperatures which give a lower yield.
Use ideas about collisions between particles in your answer.

....

....

.... **[3]**

(e) Why was the discovery of a method to make ammonia on a large scale globally important?

....

....

.... **[2]**

(f) Calculate the mass of ammonia that could be produced if 28 tonnes of nitrogen is completely converted into ammonia.
(Relative atomic masses: H = 1, N = 14.)

mass = ... tonnes **[2]**

(Total 12 marks)

5. The diagram below shows part of the Periodic Table.

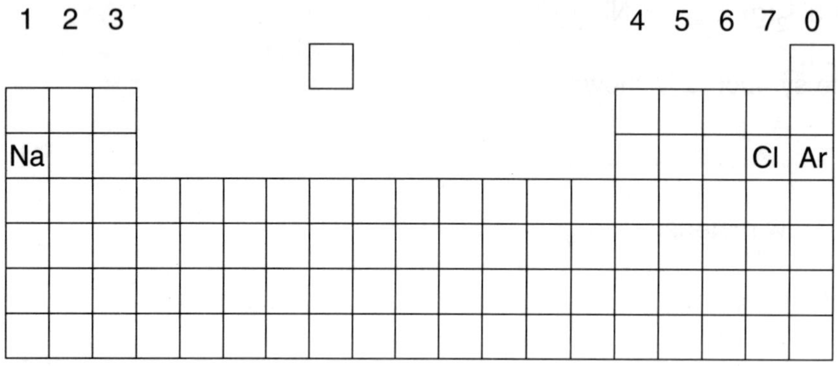

The position of three elements in the Periodic Table is shown.

(a) Describe the difference in the atomic structure of these three elements.

....

....

....

.... **[2]**

(b) Using these three elements as examples, describe the trend in chemical properties across the second period of the Periodic Table.

....

....

....

.... **[3]**

(c) Use ideas about the electronic structure of the three elements to explain this trend in chemical properties.

....

....

.... **[2]**

(Total 7 marks)

Leave blank

Leave blank

6. (a) A polymer has the structure shown below.

(i) What type of polymer is this? **[1]**

(ii) Complete the following to give the structures of the two monomers

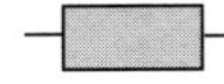

from which the above polymer could be made.

[2]

(b) Esters are frequently used as solvents in chromatography. A natural macromolecule was hydrolysed to give a mixture of amino acids. These could be identified by chromatography.

(i) What type of macromolecule was hydrolysed? **[1]**

(ii) What type of linkage was broken by hydrolysis? **[1]**

(iii) Explain why the chromatogram must be sprayed with a locating agent before the amino acids can be identified. **[1]**

(iv) Explain how is it possible to identify the amino acids from the chromatogram. **[1]**

(Total 7 marks)

7. The table gives some information about the homologous series of alkanes.

name	formula	relative molecular mass	boiling point in °C
methane	CH_4	16	–161
ethane	C_2H_6	30	**–88°C***
propane	C_3H_8	44	–42
butane	$\mathbf{C_4H_{10}}$	58	–1
pentane	C_5H_{12}	72	36

(a) Complete the table by filling in the three blank boxes. **[3]**

Leave blank

(b) Explain what is meant by the term homologous series as it applies to the alkanes.

....

....

.... **[2]**

(c) Butane exists as a number of structural isomers.

(i) What are structural isomers?

....

.... **[2]**

(ii) Draw structural (displayed) formulae for **two** structural isomers of butane.

[2]

(Total 9 marks)

8. Some coins are made of an alloy of zinc, nickel and copper. To find the percentage of zinc in the coins one coin, of mass 0.5g, was placed in 25 cm^3 of hydrochloric acid of concentration 0.5 mol/dm^3. Only the zinc reacted.

$$Zn + 2HCl \rightarrow ZnCl_2 + H_2$$

When the reaction had finished, the mixture was filtered and titrated against sodium hydroxide solution of 0.5 mol/dm^3 concentration.
To reach neutralisation point took 14.6 cm^3 of this sodium hydroxide solution.

(a) (i) How could you ascertain that the reaction between the zinc and hydrochloric acid is finished?

.... **[1]**

(ii) Explain why zinc reacted with the hydrochloric acid, but nickel and copper did not.

....

.... **[1]**

(b) (i) Calculate the volume of 0.5 mol/dm^3 hydrochloric acid which reacted with 14.6 cm^3 of 0.5 mol/dm^3 sodium hydroxide solution.

Leave blank

volume = ... cm^3 **[2]**

(ii) Calculate the volume of 0.5 mol/dm^3 hydrochloric acid, which reacted with zinc.

volume = ... cm^3 **[1]**

(iii) Calculate the mass of zinc which reacts with this volume of 0.5 mol/dm^3 hydrochloric acid.
(Relative atomic mass: Zn = 65)

mass = ... g **[3]**

(iv) What percentage of zinc was present in the coins?

percentage = ... % **[2]**

(Total 10 marks)

Model Test Answers: Paper 1

Question	Answer	Question	Answer
1	c	21	d
2	b	22	d
3	c	23	b
4	c	24	b
5	b	25	c
6	d	26	d
7	d	27	c
8	a	28	d
9	d	29	c
10	d	30	d
11	b	31	b
12	d	32	d
13	c	33	d
14	a	34	a
15	b	35	b
16	d	36	a
17	b	37	d
18	d	38	b
19	a	39	c
20	a	40	c

Model Test Answers: Paper 3-Extended

Question			Answer	Mark
1	a	i	copper(II) sulphate solution	1
		ii	copper(II) oxide	1
		iii	$CuO + H_2SO_4 \rightarrow CuSO_4 + H_2O$	
			left side	1
			right side	1
		iv	neutralisation	1

Examiner's Tip
Neutralisation reactions occur not only between acids and alkalis, but also between metal oxides and acids, as in this example, and between carbonates and acids. A salt is always formed, in this case copper(II) sulphate.

Question			Answer	Mark
	b		Filter or decant off the clear blue liquid.	1
			Heat the solution to evaporate off some of the water.	1
			Leave the remaining solution to cool.	1
			+ 1 mark for logical order in answer	1

Examiner's Tip
Crystals will form if a hot saturated solution is allowed to cool to room temperature. The excess copper(II) oxide must be removed first, then some of the water evaporated to form a saturated solution of the copper(II) sulphate.
To score the extra mark you must make three points in the correct order.

Question			Answer	Mark
	c	i	exothermic	1
		ii		

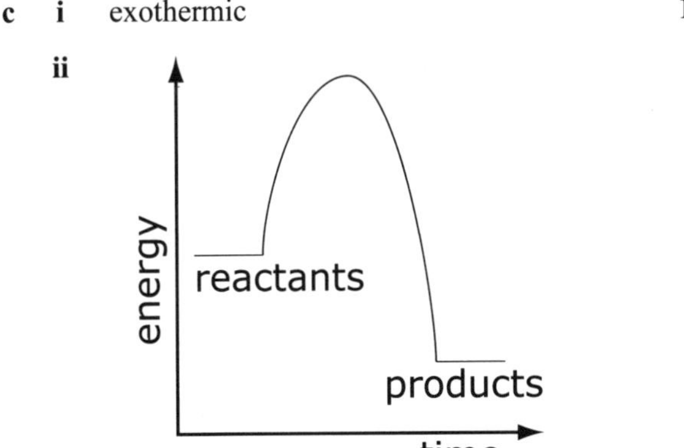

energy vs time axes drawn and labelled 1
reactants labelled at higher energy level than products 1
curve drawn to show progress of reaction 1

Examiner's Tip
Since an exothermic reaction gives out heat energy, the reactants must be at a higher energy level than the products. This is shown on the diagram.

Question			Answer	Mark
2	a	i	water	1
		ii	add to anhydrous copper(II) sulphate/cobalt chloride paper	1
			colour changes from white to blue/blue to pink	1

Examiner's Tip
The best test for water is to show that the boiling point of the liquid is 100°C, but there is not enough water in this example to do this. Anhydrous copper(II) sulphate has lost its water of crystallisation, and is white. The water restores this water of crystallisation, returning the blue colour. Always describe how to do the test, and give the colour before and after the test.

Question			Answer	Mark
	b	i	white precipitate/white cloudiness/lime water turns milky	2
		ii	The gases contain carbon dioxide.	1

Examiner's Tip
All hydrocarbons burn to give water and carbon dioxide. Lime water forms a white solid of calcium carbonate when carbon dioxide is bubbled through it. This turns the solution cloudy – a white precipitate.

Question			Answer	Mark
	c	i	$2C_8H_{18} + 25O_2 \rightarrow 18H_2O + 16CO_2$	
			Half mark each for correct formulae and balancing, rounding upto a full one mark.	1

Examiner's Tip
This is a hard equation to balance. Two molecules of octane are needed in the equation so that an even number of oxygen atoms is used. Otherwise a half molecule of oxygen would be needed. Don't be afraid of large numbers of molecules in equations – sometimes they are necessary.

Question			Answer	Mark
		ii	When the octane does not have sufficient oxygen for complete combustion	1
			some of the carbon in the octane does not combine with oxygen.	1

Examiner's Tip
Hydrocarbons only burn completely to water and carbon dioxide if there is plenty of oxygen available. In air there is not enough oxygen, so the octane does not burn completely. All of the hydrogen forms water, but some of the carbon will form carbon monoxide or carbon. The carbon gives a sooty deposit.

Question	Answer	Mark
3 a		

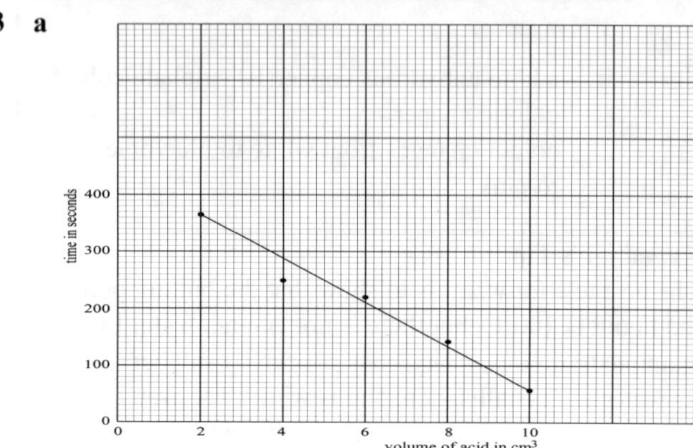

	axes	
	correctly drawn and labelled, including units	**1**
	all points plotted to + or – half a square	**1**
	a best fit line drawn ignoring the second point	**1**

Examiner's Tip
Axes need to be sensibly scaled and labelled with the thing being plotted, e.g. volume of acid, and the units, e.g. cm^3. There is just one mark for doing this correctly for both axes! Plotting of the points must be accurate. Mark them clearly with a circle or cross. The best fit line must ignore any anomalous results.

b i	This kept the total volume the same in each case,	**1**
	otherwise the concentration would not have been proportional to the volume of acid added.	**1**
ii	temperature (of the acid and water mixture)/ stirring	**1**

Examiner's Tip
By using different volumes of acid diluted with water to the same total volume each time, Sarah made sure that the concentration was proportional to the volume of acid added. This could then be plotted to give the graph. Since the rate of a reaction increases with increase in temperature, this has to be kept constant if the investigation of rate with concentration of acid is to be a fair test.

c i	4 cm^3 of acid	**1**
ii	incorrect measurement of volumes or time/not constant temperature/inconsistent stirring/ inconsistent tablets	**1**

Examiner's Tip
The result for 4 cm^3 obviously does not fit onto a straight line, which this graph should have. There are many possible reasons for this, and any sensible suggestion would score the mark in (ii).

Question	Answer	Mark
d i	Rate increases with increase in concentration.	**1**
	Rate is directly proportional to concentration.	**1**
ii	In order to react the acid particles need to collide with the solid sodium carbonate in the tablet.	**1**
	At higher concentration there are more particles of acid per cm^3,	**1**
	therefore more particles collide with the sodium carbonate each second.	**1**

Examiner's Tip
As in many questions, the number of marks indicated for each part must be carefully noted. In both (i) and (ii) it would be easy to write less than the number of points needed for full marks. The rate of a reaction depends on the number of particles which collide each second. Of these collisions a proportion will result in the formation of products. The same proportion of a larger number of collisions will result in the formation of more product in a certain length of time, i.e. a greater rate of reaction.

4 a	reversible/can go both ways/can form an equilibrium	**1**

Examiner's Tip
Any of the above answers are acceptable although saying that it shows that the reaction is reversible is probably the easiest to remember.

b	nitrogen from air	**1**
	hydrogen from crude oil/natural gas	**1**
c	Increasing the pressure increases the rate of reaction (because there are a greater number of successful collisions between the greater numbers of particles present);	**1**
	increasing the pressure increases the yield (because it pushes the equilibrium to the right and produces more ammonia).	**1**

Examiner's Tip
This is a difficult A* question. You have to use your knowledge of reaction rates to realise that increasing the pressure of gases increases their concentration and so will mean that more successful collisions will take place because there are more particles present. You also have to understand that since this equilibrium equation shows fewer molecules on the right hand side of the equation, increasing the pressure will force the equilibrium to the side with the smaller number of molecules, in this case to the right. This is often referred to as 'Le Chatelier's Principle'.

Question	Answer	Mark

d *Three from:*
Although the yield is high at low temperatures the rate of reaction is slow;
this is because gas particles have less energy and so there are fewer successful collisions;
using a higher temperature will give a lower equilibrium yield but this lower yield will be obtained much more quickly;
it is important economically to produce ammonia quickly and so higher temperatures are used which give a fast reaction rate;
a catalyst can be used to help speed up the rate at which the equilibrium is established. **3**

Examiner's Tip
Another difficult question but the mark scheme enables you to score three marks without giving every single answer on the list. The examiner will look to see that you have understood the idea that too low a temperature will give too slow a rate of reaction because of fewer successful collisions between particles.

e Ammonia is used to make artificial fertilisers; **1**
this was important because of the rapidly expanding world population requiring increased food production. **1**

f $2 \times (14 + 3)$ **1**
= 34 tonnes **1**

5 a atomic number increases from Na – Cl – Ar/increases 11 – 17 – 18 **1**
number of electrons in outer shell increases: Na – 1, Cl – 7 and Ar – 8 **1**

Examiner's Tip
It is important to write about the number of protons, since this decides the element's position in the Periodic Table, and the number of electrons in the outer shell, since this determines the chemical behaviour of the element.

b Sodium is a very reactive metal. **1**
Chlorine is a very reactive non-metal. **1**
Argon is an unreactive gas. **1**

c Sodium has one electron in its outer shell which is easily lost to get the stable electronic structure of argon – a typical metal property. **1**
Chlorine has seven electrons in its outer shell and easily gains one more to get the stable electronic structure of argon – a typical non-metal property. **1**

Examiner's Tip
These two questions look at the relationship between the number of electrons in the outer shell of an atom and its chemical properties. The fact that chemical bonding leads to each atom having a full outer electron shell, which is the same electronic structure as a noble gas, is an essential feature of chemistry.

Question	Answer	Mark

6 a i polyamide or condensation polymer **1**

ii HOOC – [ ] – COOH **2**

NH_2– [ ] – NH_2

b i protein or polypeptide or polyamide **1**

ii peptide or amide **1**

iii amino acids are colourless or to develop it **1**

iv using colour or from position

OR discussion of R_f **OR** compare with amino acids. **1**

7 a

name	formula	relative molecular mass	boiling point in °C	
methane	CH_4	16	–161	
ethane	C_2H_6	30	**–88°C***	1
propane	C_3H_8	44	–42	1
butane	$\mathbf{C_4H_{10}}$	58	–1	1
pentane	C_5H_{12}	72	36	

* *(any negative value between –70 and –130 accepted)*

Examiner's Tip
Look carefully at the information in the table and use it to help fill in the blanks. You will see that each set of information has a pattern. Use the pattern to work out the correct value for the blank box.

b A homologous series is a series of compounds each differing from the last by the same group of atoms or each having the same general formula. **1**
Each alkane has CH_2 more than the one before
or
the general formula C_nH_{2n+2}. **1**

Examiner's Tip
The question says 'as it applies to the alkanes' so your answer must say exactly how the term does apply to the alkanes.

Question	Answer	Mark
c i	Structural isomers have the same molecular formula	1
	but different structural (displayed) formulae.	1
ii	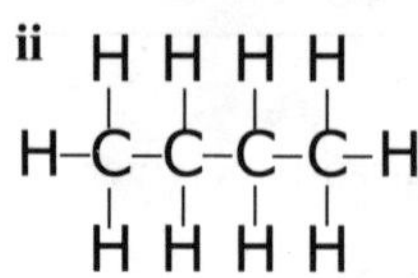	
	one mark for each diagram	2

Examiner's Tip
It is easy to draw a straight chain alkane with one of the carbon atoms pointing up or down and think this is an isomer. To be different the structural formula must actually have a carbon atom joined on in a different place.

Question	Answer	Mark
8 a i	no more bubbles/effervescence/hydrogen given off	1
ii	Zinc is higher in the reactivity series/ more reactive.	1
b i	$HCl + NaOH \rightarrow NaCl + H_2O$	
	mole ratio is 1 mole NaOH reacting with 1 mole HCl	1
	so 14.6 cm^3 0.5 mol/dm^3 NaOH reacts with 14.6 cm^3 0.5 mol/dm^3 HCl	1

Examiner's Tip
It is important to write the equation between hydrochloric acid and sodium hydroxide to find out that the mole ratio is 1:1. If you simply assume this you will lose a mark. Since the two solutions have the same concentration they will react in equal volumes.

Question	Answer	Mark
ii	volume of 0.5 mol/dm^3 HCl reacting with zinc = 25 – 14.6 = 10.4 cm^3	1

Examiner's Tip
This is a simple subtraction of the volume of acid used from the volume used originally to react with the zinc in the coin.

Question	Answer	Mark
iii	mole ratio from equation is 1 mole Zn to 2 moles HCl	
	10.4 cm^3 0.5 mol/dm^3 HCl contains	
	$0.5 \times \frac{10.4}{1000} = 0.0052$ moles	1
	moles Zn reacted = 0.5×0.0052 = 0.0026 moles	1
	mass Zn reacted = $0.0026 \times 65 = 0.169$ g	1

Examiner's Tip
You need to look back to the equation to see what the mole ratio of zinc to hydrochloric acid is. It is 1:2, so if you used 1:1 by mistake you would lose one mark. The volume of HCl can be used to calculate the moles of HCl, which then must be halved to get the moles of Zn.
Finally the moles of Zn must be multiplied by the relative atomic mass of zinc to get the mass in g.

Question	Answer	Mark
iv	% zinc in coins = $100 \times \frac{0.169}{0.5}$	1
	= 33.8 %	1

Examiner's Tip
The final stage is simply to divide the mass of zinc by the mass of the coin and multiply by 100 to get the %. Many candidates forget to multiply by 100.

Index